P9-API-637

Grades 1-3

Math Boosters

(Addition & Subtraction)

How to use this book

1. **Let's get started!** Start by writing the date on the top of each page so you can track your progress.
2. **Let's go!** Start with Step 1 and review what you already know.
3. **Don't forget!** Read the "Don't Forget" boxes which contain helpful explanations and examples.
4. **Let's work!** Solve the problems in numerical order step-by-step. Look at the samples problems for help and return to the "Review" box if you need to refresh your knowledge.
5. **Let's check your answers!** After you have finished solving the problems, check your answers and add up your score on each page. If you don't know how to do this, ask your parent or guardian to show you.
6. **Let's get it right!** Once you have finished checking your answers, review any errors to see where you made a mistake and then try again.

To parents

This workbook is designed for children to complete by themselves. By checking their answers and correcting errors on their own, children can strengthen their independence and develop into self-motivated learners.

At Kumon, we believe that each child should do work according to his or her ability, rather than his or her age or grade level. So, if this workbook is too difficult or too easy for your child, please choose another Kumon Math Workbook with an appropriate level of difficulty.

Math Boosters   Grades 1-3 Addition & Subtraction

● Table of Contents ●

Mental Math Addition

Adding 1

Date   /   /

Score   /100

Write the number that comes next.

(1) 1 ⟶ ☐ 2

(3) 3 ⟶ ☐ 4

(5) 5 ⟶ ☐ 6

(2) 2 ⟶ ☐ 3

(4) 4 ⟶ ☐ 5

(6) 7 ⟶ ☐ 8

1 **Add.**

2 points per question

Example

$$3 + 1 = 4$$

Three plus one equals four.

Adding 1 makes a number go forward 1.

0 1 2 ③ 4 5 6 7 8 9 10 11 12

3 + 1 is 4. 4 is one more than 3.

(1) 1 + 1 = ☐ 2

(4) 4 + 1 = ☐ 5

(7) 7 + 1 = ☐ 8

(2) 2 + 1 = ☐ 3

(5) 5 + 1 = ☐ 6

(8) 8 + 1 = ☐ 9

(3) 3 + 1 = ☐ 4

(6) 6 + 1 = ☐ 7

(9) 9 + 1 = ☐ 10

STEP 1-14
Mental Math
Addition

STEP 15-23
Mental Math
Subtraction

STEP 24-33
2-Digits Additon

STEP 34-42
3-Digits Additon

STEP 43-54
Subtraction in
Vertical Form

2　Add.

2 points per question

(1)　1 + 1 = $\boxed{2}$

(2)　3 + 1 = $\boxed{4}$

(3)　5 + 1 = $\boxed{6}$

(4)　7 + 1 = $\boxed{8}$

(5)　4 + 1 = $\boxed{5}$

(6)　2 + 1 = $\boxed{3}$

(7)　8 + 1 = $\boxed{9}$

(8)　6 + 1 = $\boxed{7}$

(9)　9 + 1 = $\boxed{10}$

3　Add.

4 points per question

(1)　4 + 1 = $\boxed{5}$

(2)　7 + 1 = $\boxed{8}$

(3)　9 + 1 = $\boxed{10}$

(4)　6 + 1 = $\boxed{7}$

(5)　2 + 1 = $\boxed{3}$

(6)　5 + 1 = $\boxed{6}$

(7)　3 + 1 = $\boxed{4}$

(8)　1 + 1 = $\boxed{2}$

(9)　8 + 1 = $\boxed{9}$

(10)　9 + 1 = $\boxed{10}$

(11)　2 + 1 = $\boxed{3}$

(12)　6 + 1 = $\boxed{7}$

(13)　7 + 1 = $\boxed{8}$

(14)　4 + 1 = $\boxed{5}$

(15)　9 + 1 = $\boxed{10}$

(16)　8 + 1 = $\boxed{9}$

Adding 2

Review STEP 1

Add.

(1) 7 + 1 = 8

(3) 5 + 1 = 6

(5) 4 + 1 = 5

(2) 6 + 1 = 7

(4) 3 + 1 = 4

(6) 8 + 1 = 9

1 Add.

2 points per question

Example

3 + 2 = 5

Three plus two equals five.

Adding 2 makes a number go forward 2.

0 1 2 ③ 4 5 6 7 8 9 10 11 12

3 + 2 is 5. 5 is two more than 3.

(1) 1 + 2 = 3

(4) 4 + 2 = 6

(7) 7 + 2 = 9

(2) 2 + 2 = 4

(5) 5 + 2 = 7

(8) 8 + 2 = 10

(3) 3 + 2 = 5

(6) 6 + 2 = 8

(9) 9 + 2 = 11

2 Add.

2 points per question

(1) $4 + 2 = 6$

(2) $3 + 2 = 5$

(3) $2 + 2 = 4$

(4) $1 + 2 = 3$

(5) $9 + 2 = 11$

(6) $8 + 2 = 10$

(7) $7 + 2 = 9$

(8) $6 + 2 = 8$

(9) $5 + 2 = 7$

3 Add.

4 points per question

(1) $4 + 2 = 6$

(2) $7 + 2 = 9$

(3) $9 + 2 = 11$

(4) $6 + 2 = 8$

(5) $2 + 2 = 4$

(6) $5 + 2 = 7$

(7) $3 + 2 = 5$

(8) $1 + 2 = 3$

(9) $8 + 2 = 10$

(10) $9 + 2 = 11$

(11) $2 + 2 = 4$

(12) $6 + 2 = 8$

(13) $7 + 2 = 9$

(14) $4 + 2 = 6$

(15) $9 + 2 = 11$

(16) $8 + 2 = 10$

Adding 3

Review STEP 2

Add.

(1) $6 + 2 = \boxed{8}$ (3) $9 + 2 = \boxed{11}$ (5) $4 + 2 = \boxed{6}$

(2) $5 + 2 = \boxed{7}$ (4) $8 + 2 = \boxed{10}$ (6) $7 + 2 = \boxed{9}$

1 Add.

2 points per question

Example

$4 + 3 = 7$

Four plus three equals seven.

Adding 3 makes a number go forward 3.

0 1 2 3 ④ 5 6 7 8 9 10 11 12

4 + 3 is 7. 7 is three more than 4.

(1) $1 + 3 = \boxed{4}$ (4) $4 + 3 = \boxed{7}$ (7) $7 + 3 = \boxed{10}$

(2) $2 + 3 = \boxed{5}$ (5) $5 + 3 = \boxed{8}$ (8) $8 + 3 = \boxed{11}$

(3) $3 + 3 = \boxed{6}$ (6) $6 + 3 = \boxed{9}$ (9) $9 + 3 = \boxed{12}$

STEP 1-14
Mental Math
Addition

STEP 15-23
Mental Math
Subtraction

STEP 24-33
2-Digits Additon

STEP 34-42
3-Digits Additon

STEP 43-54
Subtraction in
Vertical Form

2 Add.

2 points per question

(1) $4 + 3 = \boxed{7}$

(4) $1 + 3 = \boxed{4}$

(7) $6 + 3 = \boxed{9}$

(2) $5 + 3 = \boxed{8}$

(5) $2 + 3 = \boxed{5}$

(8) $7 + 3 = \boxed{10}$

(3) $6 + 3 = \boxed{9}$

(6) $3 + 3 = \boxed{6}$

(9) $8 + 3 = \boxed{11}$

3 Add.

4 points per question

(1) $4 + 3 = \boxed{7}$

(7) $3 + 3 = \boxed{6}$

(13) $7 + 3 = \boxed{16}$

(2) $7 + 3 = \boxed{10}$

(8) $1 + 3 = \boxed{4}$

(14) $4 + 3 = \boxed{7}$

(3) $9 + 3 = \boxed{12}$

(9) $8 + 3 = \boxed{11}$

(15) $9 + 3 = \boxed{12}$

(4) $6 + 3 = \boxed{9}$

(10) $9 + 3 = \boxed{9}$

(16) $8 + 3 = \boxed{11}$

(5) $2 + 3 = \boxed{5}$

(11) $2 + 3 = \boxed{5}$

(6) $5 + 3 = \boxed{8}$

(12) $6 + 3 = \boxed{9}$

Adding 4

Review STEP 2 STEP 3

Add.

(1) $3 + 2 = \boxed{5}$

(3) $4 + 2 = \boxed{6}$

(5) $6 + 3 = \boxed{9}$

(2) $5 + 3 = \boxed{8}$

(4) $9 + 2 = \boxed{11}$

(6) $8 + 3 = \boxed{11}$

1 Add.

2 points per question

Example

$5 + 4 = 9$

Five plus four equals nine.

$5 + 2 = 7$
$5 + 3 = 8$

(1) $2 + 3 = \boxed{5}$

(4) $4 + 3 = \boxed{7}$

(7) $5 + 4 = \boxed{9}$

(2) $2 + 4 = \boxed{6}$

(5) $4 + 4 = \boxed{8}$

(8) $3 + 3 = \boxed{6}$

(3) $1 + 4 = \boxed{5}$

(6) $5 + 3 = \boxed{8}$

(9) $3 + 4 = \boxed{7}$

STEP 1-14
Mental Math
Addition

STEP 15-23
Mental Math
Subtraction

STEP 24-33
2-Digits Additon

STEP 34-42
3-Digits Additon

STEP 43-54
Subtraction in
Vertical Form

2 **Add.**

2 points per question

(1) $5 + 4 = \boxed{9}$

(2) $6 + 3 = \boxed{9}$

(3) $6 + 4 = \boxed{10}$

(4) $8 + 3 = \boxed{11}$

(5) $8 + 4 = \boxed{12}$

(6) $7 + 3 = \boxed{10}$

(7) $7 + 4 = \boxed{11}$

(8) $9 + 3 = \boxed{12}$

(9) $9 + 4 = \boxed{13}$

3 **Add.**

4 points per question

(1) $7 + 4 = \boxed{11}$

(2) $4 + 4 = \boxed{8}$

(3) $8 + 4 = \boxed{12}$

(4) $6 + 4 = \boxed{16}$

(5) $3 + 4 = \boxed{7}$

(6) $1 + 4 = \boxed{5}$

(7) $5 + 4 = \boxed{9}$

(8) $9 + 4 = \boxed{13}$

(9) $7 + 4 = \boxed{11}$

(10) $2 + 4 = \boxed{6}$

(11) $5 + 4 = \boxed{9}$

(12) $8 + 4 = \boxed{12}$

(13) $4 + 4 = \boxed{8}$

(14) $2 + 4 = \boxed{6}$

(15) $9 + 4 = \boxed{13}$

(16) $3 + 4 = \boxed{7}$

Review STEP 4

Add.

(1) $3 + 4 = \boxed{7}$

(3) $4 + 4 = \boxed{8}$

(5) $6 + 4 = \boxed{10}$

(2) $2 + 4 = \boxed{6}$

(4) $5 + 4 = \boxed{9}$

(6) $8 + 4 = \boxed{12}$

---

1 Add.

2 points per question

Example

$5 + 5 = 10$

Five plus five equals ten.

$5 + 2 = 7$
$5 + 3 = 8$
$5 + 4 = 9$

(1) $2 + 4 = \boxed{6}$

(4) $4 + 4 = \boxed{8}$

(7) $5 + 5 = \boxed{10}$

(2) $2 + 5 = \boxed{7}$

(5) $4 + 5 = \boxed{9}$

(8) $3 + 4 = \boxed{7}$

(3) $1 + 5 = \boxed{6}$

(6) $5 + 4 = \boxed{9}$

(9) $3 + 5 = \boxed{8}$

STEP 1-14
Mental Math
Addition

STEP 15-23
Mental Math
Subtraction

STEP 24-33
2-Digits Additon

STEP 34-42
3-Digits Additon

STEP 43-54
Subtraction in
Vertical Form

2 Add.

2 points per question

(1) $5 + 5 = \boxed{10}$

(4) $8 + 4 = \boxed{12}$

(7) $7 + 5 = \boxed{12}$

(2) $6 + 4 = \boxed{10}$

(5) $8 + 5 = \boxed{13}$

(8) $9 + 4 = \boxed{13}$

(3) $6 + 5 = \boxed{11}$

(6) $7 + 4 = \boxed{11}$

(9) $9 + 5 = \boxed{14}$

3 Add.

4 points per question

(1) $6 + 5 = \boxed{11}$

(7) $9 + 5 = \boxed{14}$

(13) $6 + 5 = \boxed{11}$

(2) $4 + 5 = \boxed{9}$

(8) $6 + 5 = \boxed{11}$

(14) $9 + 5 = \boxed{14}$

(3) $2 + 5 = \boxed{7}$

(9) $8 + 5 = \boxed{13}$

(15) $8 + 5 = \boxed{13}$

(4) $7 + 5 = \boxed{12}$

(10) $3 + 5 = \boxed{9}$

(16) $3 + 5 = \boxed{8}$

(5) $1 + 5 = \boxed{6}$

(11) $2 + 5 = \boxed{7}$

(6) $5 + 5 = \boxed{10}$

(12) $7 + 5 = \boxed{12}$

Mental Math Addition +1 to +5

Date / /

Score   /100

Review STEP 1 **Add.**

2 points per question

(1) $2 + 1 = \boxed{3}$

(2) $3 + 1 = \boxed{4}$

(3) $5 + 1 = \boxed{6}$

(4) $6 + 1 = \boxed{7}$

(5) $8 + 1 = \boxed{9}$

(6) $9 + 1 = \boxed{10}$

Review STEP 2 **Add.**

2 points per question

(1) $2 + 2 = \boxed{4}$

(2) $3 + 2 = \boxed{5}$

(3) $4 + 2 = \boxed{6}$

(4) $6 + 2 = \boxed{8}$

(5) $5 + 2 = \boxed{7}$

(6) $7 + 2 = \boxed{9}$

(7) $9 + 2 = \boxed{11}$

(8) $8 + 2 = \boxed{10}$

Review STEP 3 **Add.**

2 points per question

(1) $1 + 3 = \boxed{4}$

(2) $3 + 3 = \boxed{6}$

(3) $2 + 3 = \boxed{5}$

(4) $5 + 3 = \boxed{8}$

(5) $6 + 3 = \boxed{9}$

(6) $4 + 3 = \boxed{7}$

(7) $7 + 3 = \boxed{10}$

(8) $9 + 3 = \boxed{12}$

(9) $8 + 3 = \boxed{11}$

11/11/23

Review STEP 4 **Add.**

3 points per question

(1) $2 + 4 = \boxed{6}$

(2) $3 + 4 = \boxed{7}$

(3) $1 + 4 = \boxed{5}$

(4) $4 + 4 = \boxed{8}$

(5) $5 + 4 = \boxed{9}$

(6) $7 + 4 = \boxed{11}$

(7) $6 + 4 = \boxed{10}$

(8) $8 + 4 = \boxed{12}$

(9) $9 + 4 = \boxed{13}$

Review STEP 5 **Add.**

3 points per question

(1) $2 + 5 = \boxed{8}$

(2) $1 + 5 = \boxed{6}$

(3) $4 + 5 = \boxed{9}$

(4) $3 + 5 = \boxed{8}$

(5) $6 + 5 = \boxed{11}$

(6) $7 + 5 = \boxed{12}$

(7) $5 + 5 = \boxed{10}$

(8) $8 + 5 = \boxed{13}$

(9) $9 + 5 = \boxed{14}$

Adding 6

Date / /

Score /100

Review STEP 5

Add.

(1) $3 + 5 =$ ☐

(2) $2 + 5 =$ ☐

(3) $4 + 5 =$ ☐

(4) $5 + 5 =$ ☐

(5) $7 + 5 =$ ☐

(6) $8 + 5 =$ ☐

1 Add.

2 points per question

Example

$5 + 6 = 11$

Five plus six equals eleven.

$5 + 3 = 8$
$5 + 4 = 9$
$5 + 5 = 10$

(1) $2 + 5 =$ ☐

(2) $2 + 6 =$ ☐

(3) $1 + 6 =$ ☐

(4) $4 + 5 =$ ☐

(5) $4 + 6 =$ ☐

(6) $5 + 5 =$ ☐

(7) $5 + 6 =$ ☐

(8) $3 + 5 =$ ☐

(9) $3 + 6 =$ ☐

2 Add.

2 points per question

(1) $5 + 6 =$ ☐

(2) $6 + 5 =$ ☐

(3) $6 + 6 =$ ☐

(4) $8 + 5 =$ ☐

(5) $8 + 6 =$ ☐

(6) $7 + 5 =$ ☐

(7) $7 + 6 =$ ☐

(8) $9 + 5 =$ ☐

(9) $9 + 6 =$ ☐

3 Add.

4 points per question

(1) $4 + 6 =$ ☐

(2) $7 + 6 =$ ☐

(3) $1 + 6 =$ ☐

(4) $5 + 6 =$ ☐

(5) $9 + 6 =$ ☐

(6) $3 + 6 =$ ☐

(7) $8 + 6 =$ ☐

(8) $2 + 6 =$ ☐

(9) $6 + 6 =$ ☐

(10) $7 + 6 =$ ☐

(11) $5 + 6 =$ ☐

(12) $4 + 6 =$ ☐

(13) $1 + 6 =$ ☐

(14) $6 + 6 =$ ☐

(15) $8 + 6 =$ ☐

(16) $3 + 6 =$ ☐

Adding 7

Review STEP 6

Add.

(1) $3 + 6 =$ ☐

(2) $2 + 6 =$ ☐

(3) $4 + 6 =$ ☐

(4) $5 + 6 =$ ☐

(5) $6 + 6 =$ ☐

(6) $9 + 6 =$ ☐

1 Add.

2 points per question

Example

$$5 + 7 = 12$$
Five plus seven equals twelve.

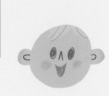

$5 + 4 = 9$
$5 + 5 = 10$
$5 + 6 = 11$

(1) $2 + 6 =$ ☐

(2) $2 + 7 =$ ☐

(3) $1 + 7 =$ ☐

(4) $4 + 6 =$ ☐

(5) $4 + 7 =$ ☐

(6) $5 + 6 =$ ☐

(7) $5 + 7 =$ ☐

(8) $3 + 6 =$ ☐

(9) $3 + 7 =$ ☐

2 **Add.**

2 points per question

(1) $5 + 7 =$

(2) $6 + 6 =$

(3) $6 + 7 =$

(4) $8 + 6 =$

(5) $8 + 7 =$

(6) $7 + 6 =$

(7) $7 + 7 =$

(8) $9 + 6 =$

(9) $9 + 7 =$

3 **Add.**

4 points per question

(1) $4 + 7 =$

(2) $7 + 7 =$

(3) $1 + 7 =$

(4) $5 + 7 =$

(5) $9 + 7 =$

(6) $3 + 7 =$

(7) $8 + 7 =$

(8) $2 + 7 =$

(9) $4 + 7 =$

(10) $5 + 7 =$

(11) $3 + 7 =$

(12) $6 + 7 =$

(13) $8 + 7 =$

(14) $9 + 7 =$

(15) $7 + 7 =$

(16) $2 + 7 =$

Adding 8

Add.

(1) $3 + 7 =$ ☐

(2) $2 + 7 =$ ☐

(3) $4 + 7 =$ ☐

(4) $5 + 7 =$ ☐

(5) $7 + 7 =$ ☐

(6) $9 + 7 =$ ☐

---

1 **Add.**

2 points per question

Example

$$5 + 8 = 13$$
Five plus eight equals thirteen.

$5 + 5 = 10$
$5 + 6 = 11$
$5 + 7 = 12$

(1) $2 + 7 =$ ☐

(2) $2 + 8 =$ ☐

(3) $1 + 8 =$ ☐

(4) $4 + 7 =$ ☐

(5) $4 + 8 =$ ☐

(6) $5 + 7 =$ ☐

(7) $5 + 8 =$ ☐

(8) $3 + 7 =$ ☐

(9) $3 + 8 =$ ☐

STEP 1-14
Mental Math Addition

STEP 15-23
Mental Math Subtraction

STEP 24-33
2-Digits Additon

STEP 34-42
3-Digits Additon

STEP 43-54
Subtraction in Vertical Form

2 **Add.**

2 points per question

(1) $5 + 8 =$

(2) $6 + 7 =$

(3) $6 + 8 =$

(4) $7 + 7 =$

(5) $7 + 8 =$

(6) $8 + 7 =$

(7) $8 + 8 =$

(8) $9 + 7 =$

(9) $9 + 8 =$

3 **Add.**

4 points per question

(1) $4 + 8 =$

(2) $7 + 8 =$

(3) $1 + 8 =$

(4) $3 + 8 =$

(5) $9 + 8 =$

(6) $8 + 8 =$

(7) $2 + 8 =$

(8) $6 + 8 =$

(9) $4 + 8 =$

(10) $7 + 8 =$

(11) $8 + 8 =$

(12) $5 + 8 =$

(13) $3 + 8 =$

(14) $9 + 8 =$

(15) $2 + 8 =$

(16) $6 + 8 =$

Adding 9

Review STEP 8

Add.

(1) $3 + 8 =$ ☐

(2) $2 + 8 =$ ☐

(3) $4 + 8 =$ ☐

(4) $5 + 8 =$ ☐

(5) $6 + 8 =$ ☐

(6) $8 + 8 =$ ☐

1 Add.

2 points per question

Example

$$5 + 9 = 14$$

Five plus nine equals fourteen.

$5 + 6 = 11$
$5 + 7 = 12$
$5 + 8 = 13$

(1) $2 + 8 =$ ☐

(2) $2 + 9 =$ ☐

(3) $1 + 9 =$ ☐

(4) $4 + 8 =$ ☐

(5) $4 + 9 =$ ☐

(6) $5 + 8 =$ ☐

(7) $5 + 9 =$ ☐

(8) $3 + 8 =$ ☐

(9) $3 + 9 =$ ☐

STEP 1-14
Mental Math
Addition

STEP 15-23
Mental Math
Subtraction

STEP 24-33
2-Digits Additon

STEP 34-42
3-Digits Additon

STEP 43-54
Subtraction in
Vertical Form

2 Add.

2 points per question

(1) $5 + 9 =$ ☐

(2) $6 + 8 =$ ☐

(3) $6 + 9 =$ ☐

(4) $8 + 8 =$ ☐

(5) $8 + 9 =$ ☐

(6) $7 + 8 =$ ☐

(7) $7 + 9 =$ ☐

(8) $9 + 8 =$ ☐

(9) $9 + 9 =$ ☐

3 Add.

4 points per question

(1) $4 + 9 =$ ☐

(2) $7 + 9 =$ ☐

(3) $5 + 9 =$ ☐

(4) $3 + 9 =$ ☐

(5) $2 + 9 =$ ☐

(6) $8 + 9 =$ ☐

(7) $6 + 9 =$ ☐

(8) $5 + 9 =$ ☐

(9) $9 + 9 =$ ☐

(10) $3 + 9 =$ ☐

(11) $1 + 9 =$ ☐

(12) $7 + 9 =$ ☐

(13) $8 + 9 =$ ☐

(14) $9 + 9 =$ ☐

(15) $4 + 9 =$ ☐

(16) $6 + 9 =$ ☐

Mental Math Addition +1 to +9

Date / /

Score /100

Review STEP 1 STEP 2 **Add.**

2 points per question

(1) $2 + 1 =$ ☐

(2) $2 + 2 =$ ☐

(3) $4 + 2 =$ ☐

(4) $4 + 1 =$ ☐

(5) $9 + 1 =$ ☐

(6) $8 + 2 =$ ☐

(7) $7 + 1 =$ ☐

(8) $6 + 2 =$ ☐

(9) $5 + 2 =$ ☐

Review STEP 3 STEP 4 **Add.**

2 points per question

(1) $9 + 3 =$ ☐

(2) $8 + 4 =$ ☐

(3) $3 + 3 =$ ☐

(4) $4 + 4 =$ ☐

(5) $6 + 3 =$ ☐

(6) $7 + 4 =$ ☐

(7) $5 + 3 =$ ☐

(8) $9 + 4 =$ ☐

(9) $8 + 3 =$ ☐

Review STEP 5 **Add.**

2 points per question

(1) $4 + 5 =$ ☐

(2) $5 + 5 =$ ☐

(3) $6 + 5 =$ ☐

(4) $9 + 5 =$ ☐

(5) $7 + 5 =$ ☐

(6) $8 + 5 =$ ☐

(7) $3 + 5 =$ ☐

(8) $2 + 5 =$ ☐

(9) $1 + 5 =$ ☐

Review STEP 6 STEP 7 Add.

2 points per question

(1) $3 + 6 = \boxed{}$

(2) $4 + 7 = \boxed{}$

(3) $7 + 6 = \boxed{}$

(4) $8 + 7 = \boxed{}$

(5) $9 + 6 = \boxed{}$

(6) $5 + 7 = \boxed{}$

(7) $1 + 6 = \boxed{}$

(8) $2 + 7 = \boxed{}$

(9) $6 + 6 = \boxed{}$

(10) $7 + 7 = \boxed{}$

(11) $8 + 6 = \boxed{}$

Review STEP 8 STEP 9 Add.

2 points per question

(1) $5 + 8 = \boxed{}$

(2) $5 + 9 = \boxed{}$

(3) $8 + 8 = \boxed{}$

(4) $7 + 9 = \boxed{}$

(5) $4 + 8 = \boxed{}$

(6) $5 + 9 = \boxed{}$

(7) $6 + 8 = \boxed{}$

(8) $8 + 9 = \boxed{}$

(9) $2 + 8 = \boxed{}$

(10) $9 + 9 = \boxed{}$

(11) $7 + 8 = \boxed{}$

(12) $3 + 9 = \boxed{}$

Mental Math Addition

Adding Sums up to 12

Review STEP 8 STEP 9

Add.

(1) $2 + 8 =$

(3) $4 + 8 =$

(5) $7 + 8 =$

(2) $2 + 9 =$

(4) $4 + 9 =$

(6) $7 + 9 =$

1 Add.

3 points per question

Example

$$7 + 5 = 12 \qquad 4 + 8 = 12$$

(1) $4 + 6 =$

(4) $3 + 7 =$

(7) $2 + 9 =$

(2) $4 + 7 =$

(5) $5 + 7 =$

(8) $3 + 9 =$

(3) $5 + 6 =$

(6) $4 + 8 =$

(9) $6 + 5 =$

2 Add.

3 points per question

(1) $7 + 3 =$ ☐

(2) $5 + 5 =$ ☐

(3) $1 + 9 =$ ☐

(4) $6 + 6 =$ ☐

(5) $7 + 4 =$ ☐

(6) $8 + 3 =$ ☐

(7) $7 + 5 =$ ☐

(8) $8 + 2 =$ ☐

(9) $9 + 3 =$ ☐

(10) $8 + 4 =$ ☐

(11) $7 + 4 =$ ☐

3 Add.

4 points per question

(1) $4 + 6 =$ ☐

(2) $7 + 4 =$ ☐

(3) $7 + 5 =$ ☐

(4) $6 + 5 =$ ☐

(5) $5 + 7 =$ ☐

(6) $8 + 4 =$ ☐

(7) $7 + 3 =$ ☐

(8) $6 + 6 =$ ☐

(9) $4 + 8 =$ ☐

(10) $3 + 9 =$ ☐

STEP **11**

Mental Math Addition

Adding Sums up to 15

Date   /   /

Score

/100

Review STEP 10

Add.

(1) $8 + 2 =$ ☐

(3) $4 + 7 =$ ☐

(5) $9 + 3 =$ ☐

(2) $6 + 5 =$ ☐

(4) $7 + 5 =$ ☐

(6) $8 + 4 =$ ☐

1 **Add.**

3 points per question

Example

$$9 + 5 = 14 \qquad 8 + 7 = 15$$

(1) $7 + 3 =$ ☐

(4) $4 + 7 =$ ☐

(7) $8 + 5 =$ ☐

(2) $2 + 8 =$ ☐

(5) $4 + 8 =$ ☐

(8) $6 + 6 =$ ☐

(3) $9 + 2 =$ ☐

(6) $9 + 4 =$ ☐

(9) $7 + 6 =$ ☐

STEP 1-14
Mental Math
Addition

STEP 15-23
Mental Math
Subtraction

STEP 24-33
2-Digits Additon

STEP 34-42
3-Digits Additon

STEP 43-54
Subtraction in
Vertical Form

2 Add.

3 points per question

(1) $6 + 3 =$ ☐

(2) $8 + 2 =$ ☐

(3) $4 + 7 =$ ☐

(4) $5 + 6 =$ ☐

(5) $5 + 7 =$ ☐

(6) $7 + 7 =$ ☐

(7) $8 + 6 =$ ☐

(8) $4 + 9 =$ ☐

(9) $9 + 6 =$ ☐

(10) $6 + 8 =$ ☐

(11) $8 + 7 =$ ☐

3 Add.

4 points per question

(1) $7 + 4 =$ ☐

(2) $8 + 4 =$ ☐

(3) $9 + 4 =$ ☐

(4) $5 + 8 =$ ☐

(5) $9 + 5 =$ ☐

(6) $9 + 6 =$ ☐

(7) $5 + 9 =$ ☐

(8) $7 + 8 =$ ☐

(9) $8 + 7 =$ ☐

(10) $6 + 8 =$ ☐

Review STEP 11

Add.

(1) $9 + 3 =$ ☐

(2) $7 + 6 =$ ☐

(3) $5 + 9 =$ ☐

(4) $8 + 5 =$ ☐

(5) $8 + 7 =$ ☐

(6) $6 + 9 =$ ☐

1 Add.

3 points per question

Example

$$9 + 7 = 16 \qquad 9 + 9 = 18$$

(1) $5 + 6 =$ ☐

(2) $7 + 6 =$ ☐

(3) $4 + 9 =$ ☐

(4) $7 + 8 =$ ☐

(5) $4 + 7 =$ ☐

(6) $7 + 7 =$ ☐

(7) $9 + 6 =$ ☐

(8) $8 + 8 =$ ☐

(9) $9 + 7 =$ ☐

2 Add.

3 points per question

(1) 6 + 5 =

(2) 9 + 4 =

(3) 8 + 6 =

(4) 7 + 8 =

(5) 5 + 9 =

(6) 9 + 6 =

(7) 9 + 5 =

(8) 7 + 9 =

(9) 9 + 9 =

(10) 8 + 9 =

(11) 9 + 8 =

3 Add.

4 points per question

(1) 7 + 6 =

(2) 9 + 3 =

(3) 7 + 9 =

(4) 8 + 8 =

(5) 7 + 7 =

(6) 9 + 8 =

(7) 9 + 7 =

(8) 8 + 7 =

(9) 9 + 9 =

(10) 6 + 9 =

Mental Math Addition

2-Digits + 1-Digit

Review STEP **8** STEP **9**

Add.

(1) $2 + 8 =$ ☐

(2) $2 + 9 =$ ☐

(3) $4 + 8 =$ ☐

(4) $4 + 9 =$ ☐

(5) $7 + 8 =$ ☐

(6) $7 + 9 =$ ☐

1 Add.

2 points per question

Example

$$12 + 3 = 15$$

Add each ones place. $2+3=5$.

(1) $10 + 1 =$ ☐

(2) $11 + 2 =$ ☐

(3) $11 + 5 =$ ☐

(4) $12 + 2 =$ ☐

(5) $12 + 4 =$ ☐

(6) $12 + 7 =$ ☐

(7) $13 + 4 =$ ☐

(8) $13 + 3 =$ ☐

(9) $14 + 4 =$ ☐

2 Add.

2 points per question

(1) $15 + 3 =$ ☐

(2) $15 + 4 =$ ☐

(3) $14 + 5 =$ ☐

(4) $14 + 3 =$ ☐

(5) $16 + 3 =$ ☐

(6) $17 + 2 =$ ☐

(7) $17 + 1 =$ ☐

(8) $18 + 1 =$ ☐

(9) $16 + 2 =$ ☐

3 Add.

4 points per question

(1) $10 + 1 =$ ☐

(2) $16 + 2 =$ ☐

(3) $14 + 3 =$ ☐

(4) $12 + 4 =$ ☐

(5) $13 + 5 =$ ☐

(6) $14 + 4 =$ ☐

(7) $10 + 7 =$ ☐

(8) $11 + 8 =$ ☐

(9) $11 + 6 =$ ☐

(10) $15 + 2 =$ ☐

(11) $16 + 3 =$ ☐

(12) $12 + 2 =$ ☐

(13) $14 + 5 =$ ☐

(14) $13 + 6 =$ ☐

(15) $13 + 4 =$ ☐

(16) $12 + 7 =$ ☐

Addition of Tens

Date / /
Score /100

Review STEP 13

Add.

(1) $15 + 1 =$ ☐

(2) $16 + 2 =$ ☐

(3) $16 + 3 =$ ☐

(4) $14 + 4 =$ ☐

(5) $18 + 1 =$ ☐

(6) $15 + 4 =$ ☐

1 Add.

2 points per question

Example

$$30 + 20 = 50$$

Add each tens place.

(1) $10 + 10 =$ ☐

(2) $20 + 10 =$ ☐

(3) $30 + 30 =$ ☐

(4) $50 + 20 =$ ☐

(5) $10 + 30 =$ ☐

(6) $20 + 30 =$ ☐

STEP 1-14
Mental Math Addition

STEP 15-23
Mental Math Subtraction

STEP 24-33
2-Digits Additon

STEP 34-42
3-Digits Additon

STEP 43-54
Subtraction in Vertical Form

2 Add.

3 points per question

(1) $20 + 7 =$ ☐

(2) $30 + 1 =$ ☐

(3) $60 + 4 =$ ☐

(4) $50 + 2 =$ ☐

(5) $70 + 5 =$ ☐

(6) $40 + 8 =$ ☐

(7) $20 + 3 =$ ☐

(8) $80 + 6 =$ ☐

3 Add.

4 points per question

(1) $30 + 50 =$ ☐

(2) $40 + 50 =$ ☐

(3) $10 + 60 =$ ☐

(4) $20 + 60 =$ ☐

(5) $50 + 7 =$ ☐

(6) $10 + 80 =$ ☐

(7) $10 + 10 =$ ☐

(8) $20 + 10 =$ ☐

(9) $30 + 10 =$ ☐

(10) $20 + 6 =$ ☐

(11) $50 + 30 =$ ☐

(12) $30 + 40 =$ ☐

(13) $20 + 20 =$ ☐

(14) $30 + 8 =$ ☐

(15) $40 + 40 =$ ☐

(16) $10 + 30 =$ ☐

Mental Math Addition

Date / /

Score

/100

Review STEP 10   Add.

2 points per question

(1) $4 + 7 =$ ☐

(2) $7 + 3 =$ ☐

(3) $6 + 6 =$ ☐

(4) $2 + 8 =$ ☐

(5) $5 + 5 =$ ☐

(6) $9 + 2 =$ ☐

(7) $8 + 4 =$ ☐

(8) $5 + 6 =$ ☐

(9) $7 + 5 =$ ☐

Review STEP 11   Add.

2 points per question

(1) $4 + 6 =$ ☐

(2) $8 + 3 =$ ☐

(3) $7 + 5 =$ ☐

(4) $6 + 7 =$ ☐

(5) $8 + 6 =$ ☐

(6) $5 + 9 =$ ☐

(7) $8 + 7 =$ ☐

(8) $9 + 6 =$ ☐

(9) $7 + 8 =$ ☐

Review STEP 12   Add.

2 points per question

(1) $4 + 8 =$ ☐

(2) $6 + 7 =$ ☐

(3) $5 + 8 =$ ☐

(4) $9 + 5 =$ ☐

(5) $6 + 9 =$ ☐

(6) $8 + 8 =$ ☐

(7) $9 + 9 =$ ☐

(8) $9 + 7 =$ ☐

(9) $8 + 9 =$ ☐

Review STEP **13** **Add.**

2 points per question

(1) $10 + 4 =$ ☐

(2) $11 + 5 =$ ☐

(3) $12 + 3 =$ ☐

(4) $14 + 2 =$ ☐

(5) $15 + 4 =$ ☐

(6) $10 + 5 =$ ☐

(7) $11 + 8 =$ ☐

(8) $17 + 2 =$ ☐

(9) $15 + 2 =$ ☐

(10) $13 + 3 =$ ☐

(11) $16 + 2 =$ ☐

Review STEP **14** **Add.**

2 points per question

(1) $40 + 30 =$ ☐

(2) $50 + 30 =$ ☐

(3) $60 + 30 =$ ☐

(4) $50 + 6 =$ ☐

(5) $20 + 20 =$ ☐

(6) $50 + 20 =$ ☐

(7) $70 + 20 =$ ☐

(8) $20 + 60 =$ ☐

(9) $30 + 8 =$ ☐

(10) $40 + 50 =$ ☐

(11) $30 + 40 =$ ☐

(12) $80 + 3 =$ ☐

Mental Math Subtraction

Subtracting 1

Date / /

Score /100

Write the number that comes before.

(1) `1` ⟵ 2   (3) `3` ⟵ 4   (5) `5` ⟵ 6

(2) `2` ⟵ 3   (4) `4` ⟵ 5   (6) `7` ⟵ 8

1 Subtract.

2 points per question

Example

$5 - 1 = 4$
Five minus one equals four.

Subtracting 1 makes a number go back 1.

0 1 2 3 4 ⑤ 6 7 8 9 10 11 12

(1) $2 - 1 =$ `1`   (4) $5 - 1 =$ `4`   (7) $8 - 1 =$ `7`

(2) $3 - 1 =$ `2`   (5) $6 - 1 =$ `5`   (8) $9 - 1 =$ `8`

(3) $4 - 1 =$ `3`   (6) $7 - 1 =$ `8`   (9) $10 - 1 =$ `9`

2 **Subtract.** 2 points per question

(1) $4 - 1 = 3$

(2) $3 - 1 = 2$

(3) $2 - 1 = 1$

(4) $1 - 1 = 0$

(5) $10 - 1 = 9$

(6) $9 - 1 = 8$

(7) $8 - 1 = 7$

(8) $7 - 1 = 6$

(9) $5 - 1 = 4$

3 **Subtract.** 4 points per question

(1) $2 - 1 = 3$

(2) $4 - 1 = 3$

(3) $6 - 1 = 5$

(4) $8 - 1 = 7$

(5) $10 - 1 = 9$

(6) $3 - 1 = 2$

(7) $5 - 1 = 4$

(8) $7 - 1 = 6$

(9) $9 - 1 = 8$

(10) $1 - 1 = 0$

(11) $6 - 1 = 5$

(12) $8 - 1 = 7$

(13) $10 - 1 = 9$

(14) $5 - 1 = 4$

(15) $7 - 1 = 6$

(16) $9 - 1 = 8$

Mental Math Subtraction

Subtracting 2

Date / /

Score /100

Review STEP 15

Subtract.

(1) $4 - 1 = \boxed{3}$ (3) $5 - 1 = \boxed{4}$ (5) $9 - 1 = \boxed{8}$

(2) $3 - 1 = \boxed{2}$ (4) $7 - 1 = \boxed{6}$ (6) $10 - 1 = \boxed{9}$

1 Subtract.

2 points per question

Example

$6 - 2 = 4$

Six minus two equals four.

Subtracting 2 makes a number go back 2.

```
0   1   2   3   4   5  (6)  7   8   9   10  11  12
```

(1) $3 - 2 = \boxed{1}$ (4) $6 - 2 = \boxed{4}$ (7) $9 - 2 = \boxed{7}$

(2) $4 - 2 = \boxed{2}$ (5) $7 - 2 = \boxed{5}$ (8) $10 - 2 = \boxed{8}$

(3) $5 - 2 = \boxed{3}$ (6) $8 - 2 = \boxed{6}$ (9) $11 - 2 = \boxed{9}$

STEP 1-14
Mental Math
Addition

STEP 15-23
Mental Math
Subtraction

STEP 24-33
2-Digits Additon

STEP 34-42
3-Digits Additon

STEP 43-54
Subtraction in
Vertical Form

2 Subtract.

2 points per question

(1) $3 - 2 = \boxed{1}$

(2) $4 - 2 = \boxed{2}$

(3) $9 - 2 = \boxed{7}$

(4) $8 - 2 = \boxed{6}$

(5) $7 - 2 = \boxed{5}$

(6) $2 - 2 = \boxed{0}$

(7) $5 - 2 = \boxed{3}$

(8) $10 - 2 = \boxed{8}$

(9) $11 - 2 = \boxed{9}$

3 Subtract.

4 points per question

(1) $3 - 2 = \boxed{1}$

(2) $7 - 2 = \boxed{5}$

(3) $4 - 2 = \boxed{2}$

(4) $10 - 2 = \boxed{8}$

(5) $9 - 2 = \boxed{7}$

(6) $11 - 2 = \boxed{9}$

(7) $8 - 2 = \boxed{6}$

(8) $6 - 2 = \boxed{4}$

(9) $4 - 2 = \boxed{2}$

(10) $5 - 2 = \boxed{3}$

(11) $8 - 2 = \boxed{6}$

(12) $6 - 2 = \boxed{4}$

(13) $3 - 2 = \boxed{1}$

(14) $2 - 2 = \boxed{0}$

(15) $9 - 2 = \boxed{7}$

(16) $10 - 2 = \boxed{8}$

STEP **17**

Mental Math Subtraction

Subtracting 3

Date   /   /

Score

/100

Review STEP 16

Subtract.

(1) 4 − 2 = 4

(2) 5 − 2 = 3

(3) 7 − 2 = 5

(4) 8 − 2 = 6

(5) 10 − 2 = 8

(6) 11 − 2 = 9

1 Subtract.

2 points per question

Example

8 + 3 = 5

Eight minus three equals five.

Subtracting 3 makes a number go back 3.

0 1 2 3 4 5 6 7 8 9 10 11 12

(1) 4 − 3 = 1

(2) 5 − 3 = 2

(3) 6 − 3 = 3

(4) 7 − 3 = 4

(5) 8 − 3 = 5

(6) 9 − 3 = 6

(7) 10 − 3 = 7

(8) 11 − 3 = 8

(9) 12 − 3 = 9

2 Subtract.

2 points per question

(1) $4 - 3 = \boxed{1}$

(2) $3 - 3 = \boxed{0}$

(3) $10 - 3 = \boxed{7}$

(4) $9 - 3 = \boxed{6}$

(5) $8 - 3 = \boxed{5}$

(6) $11 - 3 = \boxed{8}$

(7) $12 - 3 = \boxed{9}$

(8) $7 - 3 = \boxed{4}$

(9) $6 - 3 = \boxed{3}$

3 Subtract.

4 points per question

(1) $5 - 3 = \boxed{2}$

(2) $4 - 3 = \boxed{1}$

(3) $6 - 3 = \boxed{6}$

(4) $9 - 3 = \boxed{6}$

(5) $7 - 3 = \boxed{4}$

(6) $10 - 3 = \boxed{7}$

(7) $8 - 3 = \boxed{5}$

(8) $11 - 3 = \boxed{8}$

(9) $9 - 3 = \boxed{6}$

(10) $12 - 3 = \boxed{9}$

(11) $6 - 3 = \boxed{3}$

(12) $5 - 3 = \boxed{2}$

(13) $7 - 3 = \boxed{4}$

(14) $10 - 3 = \boxed{7}$

(15) $3 - 3 = \boxed{0}$

(16) $4 - 3 = \boxed{1}$

Mental Math Subtraction

Subtraction −1 to −5

Date 11/11/23

Score   /100

Review STEP 17

Subtract.

(1) $5 - 3 = \boxed{2}$ (3) $6 - 3 = \boxed{3}$ (5) $10 - 3 = \boxed{7}$

(2) $7 - 3 = \boxed{4}$ (4) $9 - 3 = \boxed{6}$ (6) $12 - 3 = \boxed{9}$

1 **Subtract.**

2 points per question

Example

$$7 - 4 = 3 \qquad 7 - 5 = 2$$

Let's practice subtraction problems −1 to −5.

(1) $4 - 1 = \boxed{3}$ (5) $5 - 2 = \boxed{3}$ (9) $6 - 3 = \boxed{3}$

(2) $4 - 2 = \boxed{4}$ (6) $5 - 4 = \boxed{1}$ (10) $6 - 4 = \boxed{2}$

(3) $4 - 3 = \boxed{1}$ (7) $5 - 5 = \boxed{0}$ (11) $6 - 5 = \boxed{1}$

(4) $5 - 1 = \boxed{4}$ (8) $6 - 2 = \boxed{4}$

11/11/23

2 **Subtract.**

3 points per question

(1) $6 - 2 = \boxed{4}$

(5) $7 - 2 = \boxed{5}$

(9) $8 - 3 = \boxed{5}$

(2) $6 - 1 = \boxed{5}$

(6) $7 - 4 = \boxed{3}$

(10) $8 - 5 = \boxed{3}$

(3) $6 - 3 = \boxed{3}$

(7) $7 - 3 = \boxed{4}$

(11) $8 - 2 = \boxed{6}$

(4) $6 - 5 = \boxed{1}$

(8) $7 - 5 = \boxed{2}$

(12) $8 - 4 = \boxed{4}$

3 **Subtract.**

3 points per question

(1) $7 - 2 = \boxed{5}$

(6) $9 - 2 = \boxed{7}$

(11) $10 - 3 = \boxed{7}$

(2) $6 - 4 = \boxed{2}$

(7) $9 - 4 = \boxed{5}$

(12) $10 - 4 = \boxed{6}$

(3) $8 - 1 = \boxed{7}$

(8) $9 - 3 = \boxed{6}$

(13) $10 - 5 = \boxed{5}$

(4) $8 - 5 = \boxed{3}$

(9) $9 - 5 = \boxed{4}$

(14) $10 - 2 = \boxed{8}$

(5) $8 - 3 = \boxed{5}$

(10) $10 - 1 = \boxed{9}$

Review STEP 18

Subtract.

(1) $5 - 2 =$ 3

(2) $5 - 4 =$ 1

(3) $6 - 3 =$ 3

(4) $7 - 3 =$ 4

(5) $9 - 4 =$ 5

(6) $8 - 5 =$ 3

1 Subtract.

2 points per question

Example

$$10 - 4 = 6$$

(1) $6 - 2 =$ 4

(2) $6 - 4 =$ 2

(3) $6 - 3 =$ 3

(4) $7 - 2 =$ 5

(5) $7 - 3 =$ 4

(6) $8 - 2 =$ 6

(7) $8 - 4 =$ 4

(8) $9 - 3 =$ 6

(9) $9 - 4 =$ 5

(10) $10 - 2 =$ 8

(11) $10 - 4 =$ 6

2 **Subtract.**

3 points per question

(1) $7 - 6 = \boxed{1}$

(2) $7 - 5 = \boxed{2}$

(3) $8 - 6 = \boxed{2}$

(4) $8 - 7 = \boxed{1}$

(5) $8 - 8 = \boxed{0}$

(6) $9 - 7 = \boxed{2}$

(7) $9 - 5 = \boxed{4}$

(8) $9 - 8 = \boxed{1}$

(9) $9 - 9 = \boxed{0}$

(10) $10 - 6 = \boxed{4}$

(11) $10 - 5 = \boxed{5}$

(12) $10 - 8 = \boxed{2}$

3 **Subtract.**

3 points per question

(1) $6 - 3 = \boxed{3}$

(2) $8 - 4 = \boxed{4}$

(3) $9 - 6 = \boxed{3}$

(4) $7 - 5 = \boxed{2}$

(5) $10 - 4 = \boxed{6}$

(6) $8 - 7 = \boxed{1}$

(7) $10 - 8 = \boxed{2}$

(8) $7 - 3 = \boxed{4}$

(9) $9 - 2 = \boxed{7}$

(10) $6 - 5 = \boxed{1}$

(11) $10 - 7 = \boxed{3}$

(12) $8 - 6 = \boxed{2}$

(13) $9 - 5 = \boxed{4}$

(14) $10 - 6 = \boxed{4}$

Review STEP 19

Subtract.

(1) $8 - 3 =$ ☐

(2) $9 - 2 =$ ☐

(3) $7 - 4 =$ ☐

(4) $10 - 6 =$ ☐

(5) $9 - 5 =$ ☐

(6) $10 - 8 =$ ☐

1 Subtract.

2 points per question

Example

$$12 - 5 = 7$$

(1) $10 - 2 =$ ☐

(2) $10 - 3 =$ ☐

(3) $10 - 5 =$ ☐

(4) $10 - 4 =$ ☐

(5) $11 - 5 =$ ☐

(6) $11 - 4 =$ ☐

(7) $11 - 2 =$ ☐

(8) $11 - 3 =$ ☐

(9) $12 - 5 =$ ☐

(10) $12 - 3 =$ ☐

(11) $12 - 4 =$ ☐

2 Subtract.

3 points per question

(1) $10 - 6 =$ ☐

(2) $10 - 8 =$ ☐

(3) $10 - 7 =$ ☐

(4) $11 - 6 =$ ☐

(5) $11 - 8 =$ ☐

(6) $11 - 9 =$ ☐

(7) $12 - 7 =$ ☐

(8) $12 - 9 =$ ☐

(9) $12 - 8 =$ ☐

3 Subtract.

3 points per question

(1) $11 - 6 =$ ☐

(2) $10 - 7 =$ ☐

(3) $11 - 9 =$ ☐

(4) $12 - 8 =$ ☐

(5) $11 - 7 =$ ☐

(6) $10 - 2 =$ ☐

(7) $11 - 8 =$ ☐

(8) $10 - 6 =$ ☐

(9) $12 - 6 =$ ☐

(10) $12 - 5 =$ ☐

(11) $12 - 7 =$ ☐

(12) $10 - 8 =$ ☐

(13) $12 - 9 =$ ☐

(14) $11 - 4 =$ ☐

(15) $12 - 3 =$ ☐

(16) $12 - 4 =$ ☐

(17) $11 - 5 =$ ☐

Review STEP **20**

Subtract.

(1) $10 - 7 = \boxed{3}$ (3) $11 - 6 = \boxed{5}$ (5) $11 - 9 = \boxed{2}$

(2) $12 - 7 = \boxed{5}$ (4) $12 - 5 = \boxed{7}$ (6) $10 - 8 = \boxed{2}$

1 **Subtract.**

2 points per question

Example

$$14 - 6 = 8$$

(1) $10 - 3 = \boxed{7}$ (5) $12 - 5 = \boxed{7}$ (9) $13 - 4 = \boxed{9}$

(2) $10 - 5 = \boxed{5}$ (6) $12 - 6 = \boxed{6}$ (10) $14 - 5 = \boxed{9}$

(3) $11 - 5 = \boxed{6}$ (7) $13 - 5 = \boxed{8}$ (11) $14 - 6 = \boxed{8}$

(4) $11 - 4 = \boxed{7}$ (8) $13 - 6 = \boxed{7}$

2 **Subtract.**

3 points per question

(1) $10 - 7 =$ 3

(2) $12 - 7 =$ 5

(3) $12 - 9 =$ 3

(4) $13 - 8 =$ 5

(5) $13 - 7 =$ 6

(6) $13 - 9 =$ 4

(7) $14 - 9 =$ 5

(8) $14 - 7 =$ 7

(9) $14 - 8 =$ 6

3 **Subtract.**

3 points per question

(1) $14 - 5 =$ 9

(2) $11 - 9 =$ 2

(3) $12 - 6 =$ 6

(4) $13 - 8 =$ 5

(5) $14 - 6 =$ 8

(6) $13 - 7 =$ 6

(7) $11 - 3 =$ 8

(8) $12 - 5 =$ 7

(9) $13 - 5 =$ 8

(10) $12 - 8 =$ 4

(11) $13 - 6 =$ 7

(12) $14 - 7 =$ 7

(13) $14 - 9 =$ 5

(14) $12 - 3 =$ 9

(15) $14 - 8 =$ 6

(16) $13 - 9 =$ 4

(17) $13 - 4 =$ 9

Mental Math Subtraction

Subtracting From Numbers up to 16

Date / /

Score /100

Review STEP 21

Subtract.

(1) $14 - 9 = \boxed{5}$ (3) $14 - 5 = \boxed{9}$ (5) $13 - 5 = \boxed{8}$

(2) $13 - 4 = \boxed{9}$ (4) $12 - 8 = \boxed{4}$ (6) $14 - 6 = \boxed{8}$

1 Subtract.

2 points per question

Example

$$16 - 9 = 7$$

(1) $12 - 5 = \boxed{7}$ (5) $13 - 6 = \boxed{7}$ (9) $15 - 7 = \boxed{5}$

(2) $13 - 5 = \boxed{2}$ (6) $14 - 6 = \boxed{8}$ (10) $15 - 8 = \boxed{7}$

(3) $14 - 5 = \boxed{9}$ (7) $13 - 7 = \boxed{6}$ (11) $16 - 8 = \boxed{8}$

(4) $12 - 6 = \boxed{6}$ (8) $14 - 7 = \boxed{7}$

2 Subtract.

3 points per question

(1) $14 - 7 = 7$

(2) $14 - 8 = 6$

(3) $15 - 8 = 7$

(4) $15 - 7 = $

(5) $15 - 6 = 9$

(6) $15 - 9 = 6$

(7) $16 - 7 = 9$

(8) $16 - 8 = 8$

(9) $16 - 9 = 7$

3 Subtract.

3 points per question

(1) $14 - 9 = 5$

(2) $15 - 7 = 8$

(3) $13 - 8 = 5$

(4) $16 - 8 = 8$

(5) $14 - 6 = 8$

(6) $15 - 9 = 6$

(7) $14 - 8 = 6$

(8) $16 - 7 = 9$

(9) $13 - 7 = 6$

(10) $13 - 9 = 4$

(11) $15 - 8 = 7$

(12) $14 - 7 = 7$

(13) $15 - 6 = 9$

(14) $14 - 5 = 9$

(15) $16 - 9 = 7$

(16) $13 - 6 = 7$

(17) $12 - 8 = 4$

Mental Math Subtraction

Subtracting From Numbers up to 18

Date / /

Score /100

Review STEP 22

Subtract.

(1) $15 - 9 = \boxed{6}$

(3) $14 - 8 = \boxed{6}$

(5) $15 - 7 = \boxed{8}$

(2) $16 - 8 = \boxed{8}$

(4) $16 - 9 = \boxed{7}$

(6) $13 - 9 = \boxed{4}$

---

1 **Subtract.**

2 points per question

Example

$$17 - 8 = 9$$

(1) $13 - 6 = \boxed{7}$

(5) $15 - 7 = \boxed{8}$

(9) $17 - 8 = \boxed{9}$

(2) $14 - 6 = \boxed{8}$

(6) $16 - 7 = \boxed{9}$

(10) $17 - 9 = \boxed{8}$

(3) $15 - 6 = \boxed{9}$

(7) $15 - 8 = \boxed{7}$

(11) $18 - 9 = \boxed{9}$

(4) $14 - 7 = \boxed{7}$

(8) $16 - 8 = \boxed{8}$

STEP 1-14
Mental Math
Addition

STEP 15-23
Mental Math
Subtraction

STEP 24-33
2-Digits Additon

STEP 34-42
3-Digits Additon

STEP 43-54
Subtraction in
Vertical Form

2 Subtract.

3 points per question

(1) $15 - 7 = \boxed{8}$

(2) $15 - 8 = \boxed{7}$

(3) $15 - 9 = \boxed{6}$

(4) $16 - 8 = \boxed{8}$

(5) $16 - 9 = \boxed{7}$

(6) $16 - 7 = \boxed{9}$

(7) $17 - 8 = \boxed{9}$

(8) $17 - 9 = \boxed{8}$

(9) $18 - 9 = \boxed{9}$

3 Subtract.

3 points per question

(1) $16 - 9 = \boxed{7}$

(2) $13 - 4 = \boxed{9}$

(3) $15 - 6 = \boxed{9}$

(4) $12 - 4 = \boxed{8}$

(5) $15 - 8 = \boxed{7}$

(6) $17 - 8 = \boxed{9}$

(7) $14 - 8 = \boxed{6}$

(8) $15 - 9 = \boxed{6}$

(9) $12 - 9 = \boxed{3}$

(10) $12 - 3 = \boxed{9}$

(11) $11 - 7 = \boxed{4}$

(12) $13 - 8 = \boxed{5}$

(13) $18 - 9 = \boxed{9}$

(14) $16 - 8 = \boxed{8}$

(15) $16 - 7 = \boxed{9}$

(16) $17 - 9 = \boxed{8}$

(17) $14 - 9 = \boxed{5}$

55

Mental Math Subtraction

Date   /   /

Score

/100

Review STEP 15 – STEP 17   Subtract.

1 point per question

(1) $3 - 1 =$ ⬜

(2) $7 - 1 =$ ⬜

(3) $10 - 1 =$ ⬜

(4) $4 - 2 =$ ⬜

(5) $8 - 2 =$ ⬜

(6) $12 - 3 =$ ⬜

(7) $5 - 3 =$ ⬜

(8) $10 - 3 =$ ⬜

(9) $9 - 3 =$ ⬜

(10) $2 - 2 =$ ⬜

(11) $11 - 2 =$ ⬜

(12) $8 - 3 =$ ⬜

(13) $9 - 1 =$ ⬜

(14) $3 - 3 =$ ⬜

(15) $7 - 3 =$ ⬜

Review STEP 18   Subtract.

2 points per question

(1) $7 - 4 =$ ⬜

(2) $8 - 4 =$ ⬜

(3) $10 - 4 =$ ⬜

(4) $9 - 5 =$ ⬜

(5) $7 - 5 =$ ⬜

(6) $10 - 5 =$ ⬜

(7) $6 - 4 =$ ⬜

(8) $4 - 4 =$ ⬜

(9) $9 - 4 =$ ⬜

(10) $8 - 5 =$ ⬜

(11) $6 - 5 =$ ⬜

STEP 1-14
Mental Math
Addition

STEP 15-23
Mental Math
Subtraction

STEP 24-33
2-Digits Additon

STEP 34-42
3-Digits Additon

STEP 43-54
Subtraction in
Vertical Form

Review STEP **19**   **Subtract.**   2 points per question

(1) $7 - 2 =$ ☐

(2) $8 - 5 =$ ☐

(3) $10 - 9 =$ ☐

(4) $9 - 7 =$ ☐

(5) $10 - 6 =$ ☐

(6) $9 - 8 =$ ☐

Review STEP **20** STEP **21**   **Subtract.**   2 points per question

(1) $11 - 9 =$ ☐

(2) $12 - 9 =$ ☐

(3) $11 - 3 =$ ☐

(4) $12 - 7 =$ ☐

(5) $12 - 3 =$ ☐

(6) $12 - 8 =$ ☐

(7) $14 - 9 =$ ☐

(8) $12 - 6 =$ ☐

(9) $14 - 7 =$ ☐

(10) $14 - 5 =$ ☐

(11) $14 - 6 =$ ☐

(12) $14 - 8 =$ ☐

Review STEP **22** STEP **23**   **Subtract.**   3 points per question

(1) $15 - 9 =$ ☐

(2) $16 - 9 =$ ☐

(3) $15 - 7 =$ ☐

(4) $16 - 8 =$ ☐

(5) $16 - 7 =$ ☐

(6) $17 - 8 =$ ☐

(7) $17 - 9 =$ ☐

(8) $18 - 9 =$ ☐

(9) $15 - 8 =$ ☐

STEP 24

2-Digit Addition

2-Digits + 1-Digit 1

Date   /   /

Score

/100

Review STEP 13

Add.

(1) $10 + 1 =$ ☐

(2) $11 + 1 =$ ☐

(3) $13 + 1 =$ ☐

(4) $12 + 2 =$ ☐

(5) $14 + 2 =$ ☐

(6) $15 + 3 =$ ☐

1 **Calculate.**

3 points per question

Example

● How to calculate $16 + 3$

$$16 + 3 = 19 \implies \begin{array}{r} 16 \\ + \ 3 \\ \hline 19 \end{array}$$

Align the numbers in a same place, then calculate.

(1)
$$\begin{array}{r} 16 \\ + \ \ 2 \\ \hline 18 \end{array}$$

(2)
$$\begin{array}{r} 13 \\ + \ \ 5 \\ \hline \ \ \end{array}$$

(3)
$$\begin{array}{r} 13 \\ + \ \ 6 \\ \hline \end{array}$$

(4)
$$\begin{array}{r} 14 \\ + \ \ 5 \\ \hline \end{array}$$

(5)
$$\begin{array}{r} 12 \\ + \ \ 4 \\ \hline \end{array}$$

(6)
$$\begin{array}{r} 12 \\ + \ \ 5 \\ \hline \end{array}$$

(7)
$$\begin{array}{r} 12 \\ + \ \ 6 \\ \hline \end{array}$$

(8)
$$\begin{array}{r} 12 \\ + \ \ 7 \\ \hline \end{array}$$

STEP 1-14
Mental Math
Addition

STEP 15-23
Mental Math
Subtraction

STEP 24-33
2-Digits Additon

STEP 34-42
3-Digits Additon

STEP 43-54
Subtraction in
Vertical Form

② Calculate.

4 points per question

(1)
$$\begin{array}{r} 13 \\ +3 \\ \hline \end{array}$$

(3)
$$\begin{array}{r} 14 \\ +5 \\ \hline \end{array}$$

(5)
$$\begin{array}{r} 16 \\ +2 \\ \hline \end{array}$$

(7)
$$\begin{array}{r} 15 \\ +3 \\ \hline \end{array}$$

(2)
$$\begin{array}{r} 13 \\ +4 \\ \hline \end{array}$$

(4)
$$\begin{array}{r} 11 \\ +6 \\ \hline \end{array}$$

(6)
$$\begin{array}{r} 14 \\ +4 \\ \hline \end{array}$$

(8)
$$\begin{array}{r} 17 \\ +2 \\ \hline \end{array}$$

③ Calculate.

4 points per question

(1)
$$\begin{array}{r} 17 \\ +2 \\ \hline \end{array}$$

(4)
$$\begin{array}{r} 13 \\ +7 \\ \hline \end{array}$$

(7)
$$\begin{array}{r} 18 \\ +5 \\ \hline \end{array}$$

(10)
$$\begin{array}{r} 16 \\ +8 \\ \hline \end{array}$$

(2)
$$\begin{array}{r} 17 \\ +3 \\ \hline \end{array}$$
2☐

(5)
$$\begin{array}{r} 13 \\ +9 \\ \hline \end{array}$$

(8)
$$\begin{array}{r} 17 \\ +6 \\ \hline \end{array}$$

(11)
$$\begin{array}{r} 18 \\ +9 \\ \hline \end{array}$$

(3)
$$\begin{array}{r} 17 \\ +4 \\ \hline \end{array}$$

(6)
$$\begin{array}{r} 19 \\ +2 \\ \hline \end{array}$$

(9)
$$\begin{array}{r} 19 \\ +5 \\ \hline \end{array}$$

2-Digit Addition
2-Digits + 1-Digit 2

Review STEP 24

Calculate.

(1)
```
  13
+  5
```

(2)
```
  16
+  2
```

(3)
```
  17
+  5
```

(4)
```
  12
+  8
```

1 Calculate.

3 points per question

Example ● How to calculate 36 + 8

```
  36
+  8
────
     4
```
The ones place
6 + 8 = 14
⇨
```
  36
+  8
────
     4
```
Carry over 1 to the tens place.
```
  36
+  8
────
  4 4
```
The tens place
Write 4. Carry over 1 plus 3 is 4.

(1)
```
  19
+  5
```

(3)
```
  39
+  7
```

(5)
```
  32
+  6
```

(7)
```
  44
+  5
```

(2)
```
  29
+  6
```

(4)
```
  49
+  8
```

(6)
```
  39
+  6
```

(8)
```
  48
+  5
```

STEP 1-14
Mental Math
Addition

STEP 15-23
Mental Math
Subtraction

STEP 24-33
2-Digits Additon

STEP 34-42
3-Digits Additon

STEP 43-54
Subtraction in
Vertical Form

2 Calculate.

4 points per question

(1)
```
    2 2
  +   8
  ┌─┬─┐
  └─┴─┘
```

(2)
```
    5 3
  +   7
```

(3)
```
    4 4
  +   6
```

(4)
```
    3 5
  +   5
```

(5)
```
    7 7
  +   3
```

(6)
```
    6 6
  +   4
```

(7)
```
    5 5
  +   5
```

3 Calculate.

6 points per question

(1) 63+9

(2) 77+8

(3) 38+9

(4) 44+8

(5) 42+8

(6) 84+6

(7) 75+7

(8) 56+8

STEP 26

2-Digit Addition

2-Digits + 2-Digits 1

Date   /   /

Score

/100

Review STEP 25

Calculate.

(1)
```
   24
+   5
```

(2)
```
   29
+   7
```

(3)
```
   48
+   5
```

(4)
```
   66
+   7
```

1 Calculate.

3 points per question

Example ● How to calculate 25 + 14

```
  25
+ 14
  39
```

Calculate each place by each.

| The ones place | 5 + 4 = 9 |
| The tens place | 2 + 1 = 3 |

Align the numbers by place value before you calculate.

(1)
```
   15
+ 11
  □□
```

(3)
```
   13
+ 15
```

(5)
```
   23
+ 14
```

(7)
```
   24
+ 12
```

(2)
```
   14
+ 12
  □□
```

(4)
```
   15
+ 14
```

(6)
```
   26
+ 13
```

(8)
```
   22
+ 17
```

2 Calculate.

4 points per question

(1)
```
   11
+  32
```

(2)
```
   35
+  24
```

(3)
```
   32
+  27
```

(4)
```
   21
+  45
```

(5)
```
   34
+  21
```

(6)
```
   26
+  62
```

(7)
```
   31
+  28
```

3 Calculate.

4 points per question

(1)
```
   36
+  42
```

(2)
```
   42
+  57
```

(3)
```
   23
+  54
```

(4)
```
   23
+  53
```

(5)
```
   43
+  12
```

(6)
```
   27
+  12
```

(7)
```
   28
+  61
```

(8)
```
   31
+  53
```

(9)
```
   66
+  12
```

(10)
```
   64
+  32
```

(11)
```
   43
+  14
```

(12)
```
   75
+  22
```

Review STEP 26

Calculate.

(1)
```
  25
+ 13
```

(2)
```
  22
+ 36
```

(3)
```
  47
+ 32
```

(4)
```
  64
+ 24
```

1 Calculate.

3 points per question

Example

● How to calculate 40 + 23

```
  40
+ 23
  63
```

Calculate each place by each.

| The ones place | 0 + 3 = 3 |
| The tens place | 4 + 2 = 6 |

Be careful with numbers that have 0.

(1)
```
  38
+ 40
```

(3)
```
  20
+ 54
```

(5)
```
  50
+ 40
```

(7)
```
  30
+ 50
```

(2)
```
  26
+ 30
```

(4)
```
  60
+ 27
```

(6)
```
  20
+ 60
```

(8)
```
  40
+ 30
```

STEP 1-14
Mental Math
Addition

STEP 15-23
Mental Math
Subtraction

STEP 24-33
2-Digits Additon

STEP 34-42
3-Digits Additon

STEP 43-54
Subtraction in
Vertical Form

2 Calculate.

4 points per question

(1)
```
    4 0
  + 1 3
  ┌─┬─┐
  └─┴─┘
```

(3)
```
    2 0
  + 5 3
```

(5)
```
    3 0
  +   8
```

(7)
```
    5 0
  +   7
```

(2)
```
    3 0
  + 2 5
```

(4)
```
    5 0
  + 4 2
```

(6)
```
    6 0
  +   4
```

3 Calculate.

6 points per question

(1) 56+30

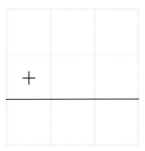

(3) 20+70

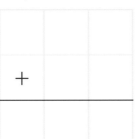

(5) 20+3

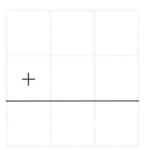

(7) 60+8

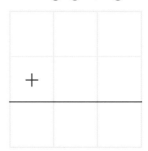

(2) 40+32

(4) 40+40

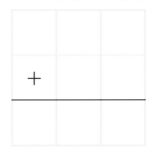

(6) 60+30

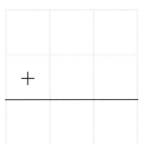

(8) 29+30

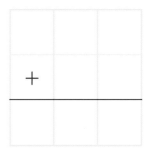

2-Digit Addition

2-Digits + 2-Digits 3

Date / /

Score /100

Review STEP 27

Calculate.

(1)
```
   47
+  30
```

(2)
```
   60
+  24
```

(3)
```
   10
+  80
```

(4)
```
   70
+   8
```

1 Calculate.

3 points per question

Example ● How to calculate 28 + 42

```
  28
+ 42
   0
```
The ones place
$8 + 2 = 10$
⟹
```
  28
+ 42
   0
```
Carry over 1 to the tens place.
```
  28
+ 42
  70
```
The tens place
Carry over 1 plus 2 is 3.
$3 + 4 = 7$

(1)
```
   25
+  25
```

(3)
```
   25
+  45
```

(5)
```
   12
+  18
```

(7)
```
   12
+  38
```

(2)
```
   25
+  35
```

(4)
```
   25
+  55
```

(6)
```
   12
+  28
```

(8)
```
   12
+  48
```

2 Calculate.

4 points per question

(1)
$$\begin{array}{r} 16 \\ + 24 \\ \hline \end{array}$$

(3)
$$\begin{array}{r} 16 \\ + 44 \\ \hline \end{array}$$

(5)
$$\begin{array}{r} 23 \\ + 17 \\ \hline \end{array}$$

(7)
$$\begin{array}{r} 35 \\ + 25 \\ \hline \end{array}$$

(2)
$$\begin{array}{r} 16 \\ + 34 \\ \hline \end{array}$$

(4)
$$\begin{array}{r} 16 \\ + 54 \\ \hline \end{array}$$

(6)
$$\begin{array}{r} 24 \\ + 27 \\ \hline \end{array}$$

(8)
$$\begin{array}{r} 39 \\ + 22 \\ \hline \end{array}$$

3 Calculate.

4 points per question

(1)
$$\begin{array}{r} 34 \\ + 16 \\ \hline \end{array}$$

(4)
$$\begin{array}{r} 52 \\ + 38 \\ \hline \end{array}$$

(7)
$$\begin{array}{r} 37 \\ + 24 \\ \hline \end{array}$$

(10)
$$\begin{array}{r} 38 \\ + 22 \\ \hline \end{array}$$

(2)
$$\begin{array}{r} 21 \\ + 49 \\ \hline \end{array}$$

(5)
$$\begin{array}{r} 45 \\ + 35 \\ \hline \end{array}$$

(8)
$$\begin{array}{r} 29 \\ + 43 \\ \hline \end{array}$$

(11)
$$\begin{array}{r} 63 \\ + 28 \\ \hline \end{array}$$

(3)
$$\begin{array}{r} 73 \\ + 17 \\ \hline \end{array}$$

(6)
$$\begin{array}{r} 12 \\ + 59 \\ \hline \end{array}$$

(9)
$$\begin{array}{r} 46 \\ + 25 \\ \hline \end{array}$$

2-Digit Addition
2-Digits + 2-Digits 4

Date / /

Score /100

Review STEP 28

Calculate.

(1)
$$\begin{array}{r} 45 \\ + 15 \\ \hline \end{array}$$

(2)
$$\begin{array}{r} 63 \\ + 17 \\ \hline \end{array}$$

(3)
$$\begin{array}{r} 29 \\ + 21 \\ \hline \end{array}$$

(4)
$$\begin{array}{r} 32 \\ + 28 \\ \hline \end{array}$$

1 Calculate.

3 points per question

Example ● How to calculate 37 + 28

$$\begin{array}{r} 37 \\ + 28 \\ \hline 5 \end{array}$$
The ones place
$7 + 8 = 15$
⇨
$$\begin{array}{r} 37 \\ + 28 \\ \hline 5 \end{array}$$
Carry over 1 to the tens place.
$$\begin{array}{r} 37 \\ + 28 \\ \hline 65 \end{array}$$
The tens place
Carried over 1 plus 3 is 4.
$4 + 2 = 6$

(1)
$$\begin{array}{r} 24 \\ + 18 \\ \hline \end{array}$$

(3)
$$\begin{array}{r} 44 \\ + 18 \\ \hline \end{array}$$

(5)
$$\begin{array}{r} 36 \\ + 16 \\ \hline \end{array}$$

(7)
$$\begin{array}{r} 36 \\ + 18 \\ \hline \end{array}$$

(2)
$$\begin{array}{r} 34 \\ + 18 \\ \hline \end{array}$$

(4)
$$\begin{array}{r} 54 \\ + 18 \\ \hline \end{array}$$

(6)
$$\begin{array}{r} 36 \\ + 17 \\ \hline \end{array}$$

(8)
$$\begin{array}{r} 36 \\ + 19 \\ \hline \end{array}$$

STEP 1-14
Mental Math
Addition

STEP 15-23
Mental Math
Subtraction

STEP 24-33
2-Digits Additon

STEP 34-42
3-Digits Additon

STEP 43-54
Subtraction in
Vertical Form

2 Calculate.

4 points per question

(1)
$$\begin{array}{r} 27 \\ + 24 \\ \hline \end{array}$$

(3)
$$\begin{array}{r} 27 \\ + 26 \\ \hline \end{array}$$

(5)
$$\begin{array}{r} 38 \\ + 16 \\ \hline \end{array}$$

(7)
$$\begin{array}{r} 38 \\ + 38 \\ \hline \end{array}$$

(2)
$$\begin{array}{r} 27 \\ + 25 \\ \hline \end{array}$$

(4)
$$\begin{array}{r} 27 \\ + 27 \\ \hline \end{array}$$

(6)
$$\begin{array}{r} 38 \\ + 27 \\ \hline \end{array}$$

(8)
$$\begin{array}{r} 38 \\ + 48 \\ \hline \end{array}$$

3 Calculate.

4 points per question

(1)
$$\begin{array}{r} 39 \\ + 24 \\ \hline \end{array}$$

(4)
$$\begin{array}{r} 75 \\ + 18 \\ \hline \end{array}$$

(7)
$$\begin{array}{r} 66 \\ + 29 \\ \hline \end{array}$$

(10)
$$\begin{array}{r} 26 \\ + 58 \\ \hline \end{array}$$

(2)
$$\begin{array}{r} 49 \\ + 34 \\ \hline \end{array}$$

(5)
$$\begin{array}{r} 28 \\ + 27 \\ \hline \end{array}$$

(8)
$$\begin{array}{r} 57 \\ + 38 \\ \hline \end{array}$$

(11)
$$\begin{array}{r} 39 \\ + 47 \\ \hline \end{array}$$

(3)
$$\begin{array}{r} 64 \\ + 19 \\ \hline \end{array}$$

(6)
$$\begin{array}{r} 49 \\ + 26 \\ \hline \end{array}$$

(9)
$$\begin{array}{r} 73 \\ + 19 \\ \hline \end{array}$$

Review STEP 29

Calculate.

(1)
```
   27
 + 36
```

(2)
```
   19
 + 45
```

(3)
```
   25
 + 56
```

(4)
```
   38
 + 34
```

1 Calculate.

3 points per question

Example ● How to calculate 63 + 56

```
   63
 + 56
 ────
    9
```
The ones place
3 + 6 = 9

⟹

```
   63
 + 56
 ────
  119
```
The tens place
6 + 5 = 11

Write 1 on the hundreds place. → ↑ ↖— Write 1 on the tens place.

Addition in the tens place leads to carrying over to the hundreds place.

(1)
```
   30
 + 67
 ────
 □□
```

(3)
```
   30
 + 87
```

(5)
```
   62
 + 46
```

(7)
```
   40
 + 60
```

(2)
```
   30
 + 77
 ────
 □□□
```

(4)
```
   32
 + 86
```

(6)
```
   72
 + 46
```

(8)
```
   40
 + 80
```

© Kumon Publishing Co., Ltd.

STEP 1-14
Mental Math Addition

STEP 15-23
Mental Math Subtraction

STEP 24-33
2-Digits Additon

STEP 34-42
3-Digits Additon

STEP 43-54
Subtraction in Vertical Form

2 Calculate.

4 points per question

(1)
$$\begin{array}{r} 82 \\ + 55 \\ \hline \end{array}$$

(4)
$$\begin{array}{r} 54 \\ + 83 \\ \hline \end{array}$$

(7)
$$\begin{array}{r} 76 \\ + 82 \\ \hline \end{array}$$

(10)
$$\begin{array}{r} 92 \\ + 50 \\ \hline \end{array}$$

(2)
$$\begin{array}{r} 73 \\ + 75 \\ \hline \end{array}$$

(5)
$$\begin{array}{r} 68 \\ + 71 \\ \hline \end{array}$$

(8)
$$\begin{array}{r} 63 \\ + 85 \\ \hline \end{array}$$

(11)
$$\begin{array}{r} 63 \\ + 75 \\ \hline \end{array}$$

(3)
$$\begin{array}{r} 67 \\ + 91 \\ \hline \end{array}$$

(6)
$$\begin{array}{r} 64 \\ + 83 \\ \hline \end{array}$$

(9)
$$\begin{array}{r} 60 \\ + 80 \\ \hline \end{array}$$

3 Calculate.

4 points per question

(1)
$$\begin{array}{r} 83 \\ + 96 \\ \hline \end{array}$$

(3)
$$\begin{array}{r} 84 \\ + 83 \\ \hline \end{array}$$

(5)
$$\begin{array}{r} 80 \\ + 80 \\ \hline \end{array}$$

(7)
$$\begin{array}{r} 90 \\ + 90 \\ \hline \end{array}$$

(2)
$$\begin{array}{r} 93 \\ + 95 \\ \hline \end{array}$$

(4)
$$\begin{array}{r} 95 \\ + 84 \\ \hline \end{array}$$

(6)
$$\begin{array}{r} 80 \\ + 90 \\ \hline \end{array}$$

(8)
$$\begin{array}{r} 86 \\ + 93 \\ \hline \end{array}$$

Review STEP 30

Calculate.

(1)
$$\begin{array}{r} 32 \\ + 70 \\ \hline \end{array}$$

(2)
$$\begin{array}{r} 32 \\ + 94 \\ \hline \end{array}$$

(3)
$$\begin{array}{r} 51 \\ + 73 \\ \hline \end{array}$$

(4)
$$\begin{array}{r} 30 \\ + 80 \\ \hline \end{array}$$

1 Calculate.

3 points per question

Example ● How to calculate $76 + 58$

Carry over
$$\begin{array}{r} 76 \\ + 58 \\ \hline 134 \end{array}$$

The ones place
$6 + 8 = 14$

The tens place
$1 + 7 + 5 = 13$

Carrying happens in both the tens place and the hundreds place.

Don't forget carry over!

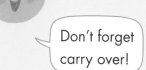

(1)
$$\begin{array}{r} 54 \\ + 56 \\ \hline \end{array}$$

(3)
$$\begin{array}{r} 56 \\ + 56 \\ \hline \end{array}$$

(5)
$$\begin{array}{r} 74 \\ + 39 \\ \hline \end{array}$$

(7)
$$\begin{array}{r} 67 \\ + 49 \\ \hline \end{array}$$

(2)
$$\begin{array}{r} 55 \\ + 56 \\ \hline \end{array}$$

(4)
$$\begin{array}{r} 75 \\ + 39 \\ \hline \end{array}$$

(6)
$$\begin{array}{r} 76 \\ + 39 \\ \hline \end{array}$$

(8)
$$\begin{array}{r} 69 \\ + 49 \\ \hline \end{array}$$

2 Calculate.

4 points per question

(1)
$$\begin{array}{r} 58 \\ + 62 \\ \hline \end{array}$$

(3)
$$\begin{array}{r} 58 \\ + 64 \\ \hline \end{array}$$

(5)
$$\begin{array}{r} 27 \\ + 96 \\ \hline \end{array}$$

(7)
$$\begin{array}{r} 39 \\ + 87 \\ \hline \end{array}$$

(2)
$$\begin{array}{r} 58 \\ + 63 \\ \hline \end{array}$$

(4)
$$\begin{array}{r} 27 \\ + 97 \\ \hline \end{array}$$

(6)
$$\begin{array}{r} 27 \\ + 98 \\ \hline \end{array}$$

(8)
$$\begin{array}{r} 49 \\ + 88 \\ \hline \end{array}$$

3 Calculate.

4 points per question

(1)
$$\begin{array}{r} 63 \\ + 67 \\ \hline \end{array}$$

(4)
$$\begin{array}{r} 35 \\ + 99 \\ \hline \end{array}$$

(7)
$$\begin{array}{r} 49 \\ + 97 \\ \hline \end{array}$$

(10)
$$\begin{array}{r} 58 \\ + 75 \\ \hline \end{array}$$

(2)
$$\begin{array}{r} 64 \\ + 67 \\ \hline \end{array}$$

(5)
$$\begin{array}{r} 35 \\ + 98 \\ \hline \end{array}$$

(8)
$$\begin{array}{r} 49 \\ + 98 \\ \hline \end{array}$$

(11)
$$\begin{array}{r} 59 \\ + 89 \\ \hline \end{array}$$

(3)
$$\begin{array}{r} 75 \\ + 67 \\ \hline \end{array}$$

(6)
$$\begin{array}{r} 38 \\ + 97 \\ \hline \end{array}$$

(9)
$$\begin{array}{r} 48 \\ + 72 \\ \hline \end{array}$$

2-Digit Addition

2-Digits + 2-Digits 7

Date / /

Score /100

Review STEP 31

Calculate.

(1) 65
   + 45

(2) 63
   + 58

(3) 48
   + 85

(4) 63
   + 78

1 **Calculate.**

3 points per question

Example ● How to calculate 86 + 97

Carry over

    86
  + 97
   183

The ones place
6 + 7 = 13

The tens place
1 + 8 + 9 = 18

When you carry over from the ones and the tens place it changes the tens and the hundreds place.

Write carried over numbers small so you don't make a mistake.

(1) 66
   + 84

(3) 66
   + 86

(5) 78
   + 75

(7) 89
   + 69

(2) 66
   + 85

(4) 78
   + 76

(6) 87
   + 68

(8) 72
   + 88

STEP 1-14
Mental Math
Addition

STEP 15-23
Mental Math
Subtraction

STEP 24-33
2-Digits Additon

STEP 34-42
3-Digits Additon

STEP 43-54
Subtraction in
Vertical Form

2 Calculate.

4 points per question

(1)
$$\begin{array}{r} 85 \\ + 85 \\ \hline \end{array}$$

(4)
$$\begin{array}{r} 96 \\ + 79 \\ \hline \end{array}$$

(7)
$$\begin{array}{r} 78 \\ + 96 \\ \hline \end{array}$$

(10)
$$\begin{array}{r} 98 \\ + 89 \\ \hline \end{array}$$

(2)
$$\begin{array}{r} 85 \\ + 86 \\ \hline \end{array}$$

(5)
$$\begin{array}{r} 97 \\ + 78 \\ \hline \end{array}$$

(8)
$$\begin{array}{r} 79 \\ + 97 \\ \hline \end{array}$$

(11)
$$\begin{array}{r} 99 \\ + 89 \\ \hline \end{array}$$

(3)
$$\begin{array}{r} 85 \\ + 87 \\ \hline \end{array}$$

(6)
$$\begin{array}{r} 98 \\ + 79 \\ \hline \end{array}$$

(9)
$$\begin{array}{r} 86 \\ + 89 \\ \hline \end{array}$$

3 Calculate.

4 points per question

(1)
$$\begin{array}{r} 85 \\ + 95 \\ \hline \end{array}$$

(3)
$$\begin{array}{r} 85 \\ + 99 \\ \hline \end{array}$$

(5)
$$\begin{array}{r} 95 \\ + 95 \\ \hline \end{array}$$

(7)
$$\begin{array}{r} 97 \\ + 96 \\ \hline \end{array}$$

(2)
$$\begin{array}{r} 85 \\ + 97 \\ \hline \end{array}$$

(4)
$$\begin{array}{r} 96 \\ + 88 \\ \hline \end{array}$$

(6)
$$\begin{array}{r} 96 \\ + 95 \\ \hline \end{array}$$

(8)
$$\begin{array}{r} 99 \\ + 99 \\ \hline \end{array}$$

2-Digit Addition

2-Digits + 2-Digits 8

Review 31 32

Calculate.

(1)
```
   74
+  56
```

(2)
```
   38
+  83
```

(3)
```
   76
+  96
```

(4)
```
   88
+  65
```

1 Calculate.

3 points per question

Example ● How to calculate 86 + 17

```
  86
+ 17
 103
```

| The ones place |
| 6 + 7 = 13 |

| The tens place |
1 + 8 + 1 = 10 ← Write 1 on the hundreds place, 0 on the tens place.

A calculation whose tens place is 0 can easily lead to a mistake. Be careful when calculating!

(1)
```
   67
+  32
```

(3)
```
   67
+  34
```
☐☐☐

(5)
```
   59
+  49
```

(7)
```
   95
+   5
```

(2)
```
   67
+  33
```
☐☐☐

(4)
```
   59
+  48
```

(6)
```
   99
+   1
```
☐☐☐

(8)
```
   98
+   7
```

STEP 1-14
Mental Math Addition

STEP 15-23
Mental Math Subtraction

STEP 24-33
2-Digits Additon

STEP 34-42
3-Digits Additon

STEP 43-54
Subtraction in Vertical Form

2 **Calculate.** 4 points per question

(1)
$$\begin{array}{r} 85 \\ + 15 \\ \hline \end{array}$$

(3)
$$\begin{array}{r} 69 \\ + 33 \\ \hline \end{array}$$

(5)
$$\begin{array}{r} 93 \\ + 7 \\ \hline \end{array}$$

(7)
$$\begin{array}{r} 37 \\ + 64 \\ \hline \end{array}$$

(2)
$$\begin{array}{r} 56 \\ + 45 \\ \hline \end{array}$$

(4)
$$\begin{array}{r} 68 \\ + 34 \\ \hline \end{array}$$

(6)
$$\begin{array}{r} 59 \\ + 45 \\ \hline \end{array}$$

(8)
$$\begin{array}{r} 63 \\ + 38 \\ \hline \end{array}$$

3 **Calculate.** 4 points per question

(1)
$$\begin{array}{r} 72 \\ + 28 \\ \hline \end{array}$$

(4)
$$\begin{array}{r} 57 \\ + 46 \\ \hline \end{array}$$

(7)
$$\begin{array}{r} 39 \\ + 67 \\ \hline \end{array}$$

(10)
$$\begin{array}{r} 95 \\ + 8 \\ \hline \end{array}$$

(2)
$$\begin{array}{r} 64 \\ + 37 \\ \hline \end{array}$$

(5)
$$\begin{array}{r} 88 \\ + 16 \\ \hline \end{array}$$

(8)
$$\begin{array}{r} 58 \\ + 49 \\ \hline \end{array}$$

(11)
$$\begin{array}{r} 76 \\ + 28 \\ \hline \end{array}$$

(3)
$$\begin{array}{r} 46 \\ + 56 \\ \hline \end{array}$$

(6)
$$\begin{array}{r} 76 \\ + 29 \\ \hline \end{array}$$

(9)
$$\begin{array}{r} 69 \\ + 39 \\ \hline \end{array}$$

2-Digit Addition

Date   /   /

Score   /100

Review STEP 24 STEP 25 Calculate.

3 points per question

(1)
$$\begin{array}{r} 16 \\ +3 \\ \hline \end{array}$$

(4)
$$\begin{array}{r} 17 \\ +3 \\ \hline \end{array}$$

(7)
$$\begin{array}{r} 27 \\ +8 \\ \hline \end{array}$$

(10)
$$\begin{array}{r} 48 \\ +7 \\ \hline \end{array}$$

(2)
$$\begin{array}{r} 14 \\ +5 \\ \hline \end{array}$$

(5)
$$\begin{array}{r} 18 \\ +4 \\ \hline \end{array}$$

(8)
$$\begin{array}{r} 39 \\ +9 \\ \hline \end{array}$$

(3)
$$\begin{array}{r} 11 \\ +8 \\ \hline \end{array}$$

(6)
$$\begin{array}{r} 16 \\ +9 \\ \hline \end{array}$$

(9)
$$\begin{array}{r} 56 \\ +5 \\ \hline \end{array}$$

Review STEP 26 STEP 27 Calculate.

2 points per question

(1)
$$\begin{array}{r} 11 \\ +13 \\ \hline \end{array}$$

(3)
$$\begin{array}{r} 43 \\ +15 \\ \hline \end{array}$$

(5)
$$\begin{array}{r} 35 \\ +54 \\ \hline \end{array}$$

(7)
$$\begin{array}{r} 50 \\ +30 \\ \hline \end{array}$$

(2)
$$\begin{array}{r} 16 \\ +11 \\ \hline \end{array}$$

(4)
$$\begin{array}{r} 25 \\ +34 \\ \hline \end{array}$$

(6)
$$\begin{array}{r} 40 \\ +27 \\ \hline \end{array}$$

(8)
$$\begin{array}{r} 28 \\ +60 \\ \hline \end{array}$$

STEP 1-14
Mental arithmetic of addition

STEP 15-23
Mental arithmetic of subtraction

STEP 24-33
Addition of two digits in a number

STEP 34-42
Addition of three digits in a number

STEP 43-54
Column subtraction

Review STEP 28 STEP 29 Calculate.

3 points per question

(1)
```
   34
+  17
```

(3)
```
   49
+  25
```

(5)
```
   24
+  36
```

(7)
```
   28
+  62
```

(2)
```
   56
+  28
```

(4)
```
   18
+  32
```

(6)
```
   39
+  47
```

(8)
```
   15
+  65
```

Review STEP 30 − STEP 33 Calculate.

3 points per question

(1)
```
   32
+  86
```

(4)
```
   78
+  45
```

(7)
```
   56
+  77
```

(10)
```
   59
+  47
```

(2)
```
   70
+  63
```

(5)
```
   89
+  53
```

(8)
```
   98
+  83
```

(3)
```
   49
+  83
```

(6)
```
   75
+  55
```

(9)
```
   73
+  28
```

3-Digit Addition

3-Digits + 3-Digits 1

Date / /

Score /100

Review STEP 30 – STEP 33

Calculate.

(1)
```
   50
+  68
```

(2)
```
   45
+  57
```

(3)
```
   34
+  96
```

(4)
```
   86
+  78
```

1 Calculate.

3 points per question

Example ● How to calculate 128 + 53

```
  128
+  53
  181
```

The ones place	$8 + 3 = 11$	→ Carry over 1 to the tens place.
The tens place	$1 + 2 + 5 = 8$	
The hundreds place	Write 1 as it makes no carry over.	

(1)
```
  115
+   7
```

(3)
```
  138
+   5
```

(5)
```
  136
+   4
```

(7)
```
  128
+  42
```

(2)
```
  128
+   3
```

(4)
```
  138
+  15
```

(6)
```
  136
+  24
```

(8)
```
  147
+  33
```

2 Calculate.

4 points per question

(1)
$$173 + 19$$

(2)
$$173 + 119$$

(3)
$$124 + 139$$

(4)
$$126 + 148$$

(5)
$$315 + 36$$

(6)
$$315 + 136$$

(7)
$$138 + 244$$

(8)
$$212 + 159$$

(9)
$$205 + 116$$

(10)
$$204 + 178$$

(11)
$$347 + 107$$

3 Calculate.

4 points per question

(1)
$$126 + 47$$

(2)
$$433 + 58$$

(3)
$$218 + 74$$

(4)
$$347 + 37$$

(5)
$$656 + 236$$

(6)
$$377 + 609$$

(7)
$$638 + 153$$

(8)
$$733 + 259$$

3-Digits + 3-Digits 2

Date / /

Score /100

Review STEP 34

Calculate.

(1)
```
  128
+  24
```

(2)
```
  163
+   8
```

(3)
```
  142
+ 219
```

(4)
```
  254
+ 108
```

1 Calculate.

3 points per question

Example ● How to calculate 372 + 253

```
  372
+ 253
  625
```

The ones place	2 + 3 = 5
The tens place	7 + 5 = 12 → Carry over 1 to the hundreds place.
The hundreds place	1 + 3 + 2 = 6

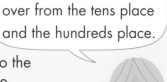

In this case you carry over from the tens place and the hundreds place.

(1)
```
  163
+  64
```

(3)
```
  163
+ 284
```

(5)
```
  256
+  52
```

(7)
```
  256
+ 272
```

(2)
```
  163
+ 174
```

(4)
```
  163
+ 294
```

(6)
```
  256
+ 162
```

(8)
```
  256
+ 282
```

2 Calculate.

4 points per question

(1)
$$157 + 52$$

(3)
$$157 + 272$$

(5)
$$184 + 33$$

(7)
$$282 + 255$$

(2)
$$157 + 162$$

(4)
$$157 + 382$$

(6)
$$184 + 143$$

(8)
$$371 + 266$$

3 Calculate.

4 points per question

(1)
$$136 + 182$$

(4)
$$357 + 81$$

(7)
$$324 + 95$$

(10)
$$178 + 81$$

(2)
$$175 + 463$$

(5)
$$147 + 372$$

(8)
$$468 + 271$$

(11)
$$654 + 294$$

(3)
$$253 + 76$$

(6)
$$286 + 153$$

(9)
$$194 + 283$$

3-Digits + 3-Digits 3

Review STEP 35

Calculate.

(1)
```
  172
+ 365
```

(2)
```
  183
+ 462
```

(3)
```
  123
+ 394
```

(4)
```
  153
+ 676
```

1 Calculate.

3 points per question

Example ● How to calculate 143 + 188

Carry over 1 to the tens place.

```
  143
+ 188
  331
```

The ones place	3 + 8 = ①1
The tens place	1 + 4 + 8 = ①3
The hundreds place	1 + 1 + 1 = 3

Carry over 1 to the hundreds place.

It leads to carry over.

(1)
```
  155
+  67
```

(2)
```
  146
+ 278
```

(3)
```
  287
+ 156
```

(4)
```
  198
+ 279
```

(5)
```
  173
+ 158
```

(6)
```
  136
+ 175
```

(7)
```
  232
+ 189
```

(8)
```
  267
+ 164
```

STEP 1-14
Mental Math
Addition

STEP 15-23
Mental Math
Subtraction

STEP 24-33
2-Digits Additon

STEP 34-42
3-Digits Additon

STEP 43-54
Subtraction in
Vertical Form

2 Calculate.

4 points per question

(1)
$$146 + 75$$

(3)
$$163 + 68$$

(5)
$$237 + 85$$

(7)
$$353 + 69$$

(2)
$$172 + 59$$

(4)
$$156 + 58$$

(6)
$$269 + 74$$

(8)
$$375 + 37$$

3 Calculate.

4 points per question

(1)
$$157 + 288$$

(4)
$$285 + 225$$

(7)
$$365 + 495$$

(10)
$$248 + 377$$

(2)
$$244 + 187$$

(5)
$$387 + 276$$

(8)
$$598 + 278$$

(11)
$$438 + 486$$

(3)
$$359 + 174$$

(6)
$$588 + 156$$

(9)
$$397 + 527$$

3-Digit Addition
3-Digits + 3-Digits 4

Date / /

Score /100

Review STEP 36

Calculate.

(1)
```
  175
+  68
```

(2)
```
  254
+  97
```

(3)
```
  235
+ 187
```

(4)
```
  435
+ 288
```

1 Calculate.

3 points per question

Example ● How to calculate 123 + 178

```
  123
+ 178
  301
```

The ones place	$3+8=11$	→ Carry over 1 to the tens place.
The tens place	$1+2+7=10$	→ Carry over 1 to the hundreds place.
The hundreds place	$1+1+1=3$	

(1)
```
  175
+ 125
```

(3)
```
  173
+ 128
```

(5)
```
  265
+ 136
```

(7)
```
  336
+ 168
```

(2)
```
  164
+ 136
```

(4)
```
  189
+ 113
```

(6)
```
  203
+ 199
```

(8)
```
  324
+ 177
```

2 Calculate.

4 points per question

(1)
$$\begin{array}{r} 153 \\ + 49 \\ \hline \end{array}$$

(3)
$$\begin{array}{r} 263 \\ + 37 \\ \hline \end{array}$$

(5)
$$\begin{array}{r} 346 \\ + 55 \\ \hline \end{array}$$

(7)
$$\begin{array}{r} 463 \\ + 38 \\ \hline \end{array}$$

(2)
$$\begin{array}{r} 175 \\ + 27 \\ \hline \end{array}$$

(4)
$$\begin{array}{r} 246 \\ + 54 \\ \hline \end{array}$$

(6)
$$\begin{array}{r} 372 \\ + 29 \\ \hline \end{array}$$

(8)
$$\begin{array}{r} 546 \\ + 54 \\ \hline \end{array}$$

3 Calculate.

4 points per question

(1)
$$\begin{array}{r} 173 \\ + 328 \\ \hline \end{array}$$

(4)
$$\begin{array}{r} 171 \\ + 429 \\ \hline \end{array}$$

(7)
$$\begin{array}{r} 345 \\ + 58 \\ \hline \end{array}$$

(10)
$$\begin{array}{r} 208 \\ + 397 \\ \hline \end{array}$$

(2)
$$\begin{array}{r} 635 \\ + 168 \\ \hline \end{array}$$

(5)
$$\begin{array}{r} 314 \\ + 487 \\ \hline \end{array}$$

(8)
$$\begin{array}{r} 627 \\ + 279 \\ \hline \end{array}$$

(11)
$$\begin{array}{r} 499 \\ + 102 \\ \hline \end{array}$$

(3)
$$\begin{array}{r} 323 \\ + 177 \\ \hline \end{array}$$

(6)
$$\begin{array}{r} 574 \\ + 27 \\ \hline \end{array}$$

(9)
$$\begin{array}{r} 397 \\ + 507 \\ \hline \end{array}$$

Review STEP 37

Calculate.

(1) 147
   + 253

(2) 343
   + 158

(3) 435
   + 268

(4) 499
   + 102

1 Calculate.

3 points per question

Example ● How to calculate 812 + 373

$$\begin{array}{r} 812 \\ + 373 \\ \hline 1185 \end{array}$$

The ones place	2 + 3 = 5
The tens place	1 + 7 = 8
The hundreds place	8 + 3 = ①1

→ Carry over 1 to the thousands place.

(1) 521
   + 613

(3) 842
   + 232

(5) 860
   + 210

(7) 971
   + 600

(2) 711
   + 333

(4) 634
   + 621

(6) 508
   + 760

(8) 746
   + 512

2 Calculate.

4 points per question

(1)
$$716 + 315$$

(3)
$$809 + 324$$

(5)
$$524 + 508$$

(7)
$$928 + 638$$

(2)
$$828 + 215$$

(4)
$$915 + 227$$

(6)
$$637 + 514$$

3 Calculate.

4 points per question

(1)
$$300 + 800$$

(4)
$$540 + 728$$

(7)
$$832 + 416$$

(10)
$$345 + 726$$

(2)
$$720 + 840$$

(5)
$$655 + 823$$

(8)
$$954 + 824$$

(11)
$$468 + 725$$

(3)
$$432 + 660$$

(6)
$$746 + 612$$

(9)
$$359 + 815$$

(12)
$$566 + 617$$

Review STEP 38

Calculate.

(1) $\begin{array}{r} 220 \\ +840 \\ \hline \end{array}$

(2) $\begin{array}{r} 851 \\ +315 \\ \hline \end{array}$

(3) $\begin{array}{r} 732 \\ +656 \\ \hline \end{array}$

(4) $\begin{array}{r} 659 \\ +817 \\ \hline \end{array}$

1 Calculate.

3 points per question

Example   ● How to calculate 583 + 641

$\begin{array}{r} 583 \\ +641 \\ \hline 1224 \end{array}$

The ones place	3 + 1 = 4
The tens place	8 + 4 = ①2	→ Carry over 1 to the hundreds place.
The hundreds place	1 + 5 + 6 = ①2	→ Carry over 1 to the thousands place.

(1) $\begin{array}{r} 591 \\ +624 \\ \hline \end{array}$

(3) $\begin{array}{r} 873 \\ +345 \\ \hline \end{array}$

(5) $\begin{array}{r} 453 \\ +762 \\ \hline \end{array}$

(7) $\begin{array}{r} 382 \\ +831 \\ \hline \end{array}$

(2) $\begin{array}{r} 672 \\ +582 \\ \hline \end{array}$

(4) $\begin{array}{r} 761 \\ +547 \\ \hline \end{array}$

(6) $\begin{array}{r} 924 \\ +391 \\ \hline \end{array}$

(8) $\begin{array}{r} 654 \\ +663 \\ \hline \end{array}$

2 Calculate.

4 points per question

(1)
$$\begin{array}{r} 244 \\ + 882 \\ \hline \end{array}$$

(3)
$$\begin{array}{r} 765 \\ + 363 \\ \hline \end{array}$$

(5)
$$\begin{array}{r} 575 \\ + 564 \\ \hline \end{array}$$

(7)
$$\begin{array}{r} 343 \\ + 863 \\ \hline \end{array}$$

(2)
$$\begin{array}{r} 129 \\ + 990 \\ \hline \end{array}$$

(4)
$$\begin{array}{r} 471 \\ + 642 \\ \hline \end{array}$$

(6)
$$\begin{array}{r} 270 \\ + 840 \\ \hline \end{array}$$

(8)
$$\begin{array}{r} 982 \\ + 222 \\ \hline \end{array}$$

3 Calculate.

4 points per question

(1)
$$\begin{array}{r} 364 \\ + 753 \\ \hline \end{array}$$

(4)
$$\begin{array}{r} 792 \\ + 754 \\ \hline \end{array}$$

(7)
$$\begin{array}{r} 743 \\ + 591 \\ \hline \end{array}$$

(10)
$$\begin{array}{r} 534 \\ + 773 \\ \hline \end{array}$$

(2)
$$\begin{array}{r} 575 \\ + 643 \\ \hline \end{array}$$

(5)
$$\begin{array}{r} 643 \\ + 985 \\ \hline \end{array}$$

(8)
$$\begin{array}{r} 682 \\ + 853 \\ \hline \end{array}$$

(11)
$$\begin{array}{r} 862 \\ + 243 \\ \hline \end{array}$$

(3)
$$\begin{array}{r} 652 \\ + 784 \\ \hline \end{array}$$

(6)
$$\begin{array}{r} 246 \\ + 863 \\ \hline \end{array}$$

(9)
$$\begin{array}{r} 874 \\ + 664 \\ \hline \end{array}$$

Review STEP 39

Calculate.

$$
\begin{array}{r} (1)\quad 870 \\ +\ 255 \\ \hline \end{array}
\qquad
\begin{array}{r} (2)\quad 762 \\ +\ 662 \\ \hline \end{array}
\qquad
\begin{array}{r} (3)\quad 463 \\ +\ 791 \\ \hline \end{array}
\qquad
\begin{array}{r} (4)\quad 684 \\ +\ 851 \\ \hline \end{array}
$$

1 Calculate.

3 points per question

Example ● How to calculate $128 + 873$

$$
\begin{array}{r} 128 \\ +873 \\ \hline 1001 \end{array}
$$

The ones place	$8+3=11$, Carry over 1 to the tens place.
The tens place	$1+2+7=10$, Carry over 1 to the hundreds place.
The hundreds place	$1+1+8=10$, Carry over 1 to the thousands place.

$$
\begin{array}{r} (1)\quad 354 \\ +\ 646 \\ \hline \end{array}
\qquad
\begin{array}{r} (3)\quad 482 \\ +\ 519 \\ \hline \end{array}
\qquad
\begin{array}{r} (5)\quad 246 \\ +\ 754 \\ \hline \end{array}
\qquad
\begin{array}{r} (7)\quad 348 \\ +\ 657 \\ \hline \end{array}
$$

$$
\begin{array}{r} (2)\quad 237 \\ +\ 763 \\ \hline \end{array}
\qquad
\begin{array}{r} (4)\quad 185 \\ +\ 817 \\ \hline \end{array}
\qquad
\begin{array}{r} (6)\quad 185 \\ +\ 818 \\ \hline \end{array}
\qquad
\begin{array}{r} (8)\quad 547 \\ +\ 464 \\ \hline \end{array}
$$

2 Calculate.

4 points per question

(1)
$$128 + 972$$

(4)
$$376 + 848$$

(7)
$$588 + 546$$

(10)
$$767 + 366$$

(2)
$$458 + 765$$

(5)
$$167 + 835$$

(8)
$$924 + 399$$

(11)
$$244 + 887$$

(3)
$$534 + 596$$

(6)
$$829 + 293$$

(9)
$$367 + 784$$

3 Calculate.

4 points per question

(1)
$$256 + 744$$

(3)
$$293 + 888$$

(5)
$$978 + 122$$

(7)
$$976 + 24$$

(2)
$$329 + 799$$

(4)
$$862 + 498$$

(6)
$$974 + 56$$

(8)
$$995 + 9$$

4-Digits + 3-Digits

Date / /

Score   /100

Review STEP 40

Calculate.

(1) 254
   + 746

(2) 477
   + 523

(3) 862
   + 498

(4) 674
   + 858

1 Calculate.

5 points per question

Example ● How to calculate 2253 + 142

      2253
    + 142
       95

The ones place
3 + 2 = 5

The tens place
5 + 4 = 9

      2253
    + 142
      2395

The hundreds place
2 + 1 = 3

Write 2 on the thousands place.

(1) 1000
   + 100

(3) 1200
   + 430

(5) 1255
   + 324

(2) 1070
   + 800

(4) 1376
   + 512

(6) 1136
   + 453

2 Calculate.

5 points per question

(1)
$$\begin{array}{r} 2425 \\ +316 \\ \hline \end{array}$$

(4)
$$\begin{array}{r} 2253 \\ +142 \\ \hline \end{array}$$

(7)
$$\begin{array}{r} 1366 \\ +437 \\ \hline \end{array}$$

(2)
$$\begin{array}{r} 2428 \\ +244 \\ \hline \end{array}$$

(5)
$$\begin{array}{r} 2351 \\ +253 \\ \hline \end{array}$$

(8)
$$\begin{array}{r} 4275 \\ +586 \\ \hline \end{array}$$

(3)
$$\begin{array}{r} 3513 \\ +268 \\ \hline \end{array}$$

(6)
$$\begin{array}{r} 1480 \\ +266 \\ \hline \end{array}$$

3 Calculate.

5 points per question

(1)
$$\begin{array}{r} 2000 \\ +200 \\ \hline \end{array}$$

(3)
$$\begin{array}{r} 2454 \\ +760 \\ \hline \end{array}$$

(5)
$$\begin{array}{r} 6940 \\ +287 \\ \hline \end{array}$$

(2)
$$\begin{array}{r} 2253 \\ +439 \\ \hline \end{array}$$

(4)
$$\begin{array}{r} 1500 \\ +764 \\ \hline \end{array}$$

(6)
$$\begin{array}{r} 4798 \\ +231 \\ \hline \end{array}$$

Review STEP 41

Calculate.

(1)
```
  1300
+  545
```

(2)
```
  3595
+  320
```

(3)
```
  1454
+  367
```

(4)
```
  2531
+  480
```

1 Calculate.

5 points per question

Example ● How to calculate 2416 + 1270

```
  2416
+ 1270
    86
```
| The ones place |
| 6 + 0 = 6 |

| The tens place |
| 1 + 7 = 8 |

```
  2416
+ 1270
  3686
```
| The hundreds place |
| 4 + 2 = 6 |

| The thousands place |
| 2 + 1 = 3 |

(1)
```
  2000
+ 1000
```

(3)
```
  3540
+ 1430
```

(5)
```
  3208
+ 1451
```

(2)
```
  1300
+ 1100
```

(4)
```
  2435
+ 1260
```

(6)
```
  1293
+ 2306
```

STEP 1-14
Mental Math Addition

STEP 15-23
Mental Math Subtraction

STEP 24-33
2-Digits Additon

STEP 34-42
3-Digits Additon

STEP 43-54
Subtraction in Vertical Form

2 Calculate.

5 points per question

(1)
$$2726 + 1248$$

(2)
$$4529 + 1334$$

(3)
$$3648 + 2105$$

(4)
$$3615 + 1267$$

(5)
$$1293 + 2385$$

(6)
$$3208 + 1455$$

(7)
$$1437 + 3285$$

(8)
$$2418 + 1298$$

3 Calculate.

5 points per question

(1)
$$1280 + 2319$$

(2)
$$3635 + 1245$$

(3)
$$4142 + 2399$$

(4)
$$1234 + 3421$$

(5)
$$1158 + 3294$$

(6)
$$2547 + 1935$$

3-Digit Addition

© Kumon Publishing Co., Ltd.

Review STEP **34** STEP **35**   **Calculate.**   2 points per question

(1)
$$
\begin{array}{r}
253 \\
+128 \\
\hline
\end{array}
$$

(3)
$$
\begin{array}{r}
146 \\
+434 \\
\hline
\end{array}
$$

(5)
$$
\begin{array}{r}
349 \\
+170 \\
\hline
\end{array}
$$

(7)
$$
\begin{array}{r}
651 \\
+290 \\
\hline
\end{array}
$$

(2)
$$
\begin{array}{r}
147 \\
+229 \\
\hline
\end{array}
$$

(4)
$$
\begin{array}{r}
325 \\
+567 \\
\hline
\end{array}
$$

(6)
$$
\begin{array}{r}
452 \\
+166 \\
\hline
\end{array}
$$

(8)
$$
\begin{array}{r}
374 \\
+233 \\
\hline
\end{array}
$$

Review STEP **36** STEP **37**   **Calculate.**   2 points per question

(1)
$$
\begin{array}{r}
345 \\
+155 \\
\hline
\end{array}
$$

(3)
$$
\begin{array}{r}
418 \\
+194 \\
\hline
\end{array}
$$

(5)
$$
\begin{array}{r}
264 \\
+178 \\
\hline
\end{array}
$$

(7)
$$
\begin{array}{r}
399 \\
+175 \\
\hline
\end{array}
$$

(2)
$$
\begin{array}{r}
147 \\
+265 \\
\hline
\end{array}
$$

(4)
$$
\begin{array}{r}
324 \\
+299 \\
\hline
\end{array}
$$

(6)
$$
\begin{array}{r}
446 \\
+387 \\
\hline
\end{array}
$$

(8)
$$
\begin{array}{r}
537 \\
+373 \\
\hline
\end{array}
$$

Review STEP 38 – STEP 40   Calculate.

4 points per question

(1)
$$
\begin{array}{r}
540 \\
+\ 730 \\
\hline
\end{array}
$$

(4)
$$
\begin{array}{r}
849 \\
+\ 429 \\
\hline
\end{array}
$$

(7)
$$
\begin{array}{r}
575 \\
+\ 648 \\
\hline
\end{array}
$$

(10)
$$
\begin{array}{r}
548 \\
+\ 796 \\
\hline
\end{array}
$$

(2)
$$
\begin{array}{r}
825 \\
+\ 551 \\
\hline
\end{array}
$$

(5)
$$
\begin{array}{r}
144 \\
+\ 872 \\
\hline
\end{array}
$$

(8)
$$
\begin{array}{r}
652 \\
+\ 789 \\
\hline
\end{array}
$$

(11)
$$
\begin{array}{r}
988 \\
+\ \ 56 \\
\hline
\end{array}
$$

(3)
$$
\begin{array}{r}
436 \\
+\ 728 \\
\hline
\end{array}
$$

(6)
$$
\begin{array}{r}
763 \\
+\ 552 \\
\hline
\end{array}
$$

(9)
$$
\begin{array}{r}
453 \\
+\ 668 \\
\hline
\end{array}
$$

Review STEP 41 STEP 42   Calculate.

4 points per question

(1)
$$
\begin{array}{r}
4374 \\
+\ \ 781 \\
\hline
\end{array}
$$

(3)
$$
\begin{array}{r}
4576 \\
+\ \ 487 \\
\hline
\end{array}
$$

(5)
$$
\begin{array}{r}
1862 \\
+\ 3579 \\
\hline
\end{array}
$$

(2)
$$
\begin{array}{r}
3615 \\
+\ \ 596 \\
\hline
\end{array}
$$

(4)
$$
\begin{array}{r}
2253 \\
+\ \ 942 \\
\hline
\end{array}
$$

(6)
$$
\begin{array}{r}
2351 \\
+\ 6439 \\
\hline
\end{array}
$$

Subtraction in Vertical Form
2-Digits - 2-Digits 1

Review STEP 23

Calculate.

(1) $14 - 8 =$ ☐

(3) $16 - 9 =$ ☐

(5) $15 - 7 =$ ☐

(2) $15 - 6 =$ ☐

(4) $18 - 9 =$ ☐

(6) $16 - 8 =$ ☐

1 Calculate.

3 points per question

Example ● How to calculate 29 − 15

$$\begin{array}{r} 29 \\ -15 \\ \hline 4 \end{array}$$
Calculate by each place value.
The ones place
$9 - 5 = 4$
⇨
$$\begin{array}{r} 29 \\ -15 \\ \hline 14 \end{array}$$
The tens place
$2 - 1 = 1$

(1)
$$\begin{array}{r} 15 \\ -12 \\ \hline \square\square \end{array}$$

(3)
$$\begin{array}{r} 26 \\ -13 \\ \hline \end{array}$$

(5)
$$\begin{array}{r} 27 \\ -16 \\ \hline \end{array}$$

(7)
$$\begin{array}{r} 35 \\ -24 \\ \hline \end{array}$$

(2)
$$\begin{array}{r} 25 \\ -12 \\ \hline \square\square \end{array}$$

(4)
$$\begin{array}{r} 27 \\ -14 \\ \hline \end{array}$$

(6)
$$\begin{array}{r} 35 \\ -23 \\ \hline \end{array}$$

(8)
$$\begin{array}{r} 39 \\ -25 \\ \hline \end{array}$$

2 Calculate.

4 points per question

(1)
$$\begin{array}{r} 47 \\ -\ 15 \\ \hline \end{array}$$

(3)
$$\begin{array}{r} 49 \\ -\ 27 \\ \hline \end{array}$$

(5)
$$\begin{array}{r} 56 \\ -\ 30 \\ \hline \end{array}$$

(7)
$$\begin{array}{r} 56 \\ -\ \ 6 \\ \hline \end{array}$$

(2)
$$\begin{array}{r} 48 \\ -\ 16 \\ \hline \end{array}$$

(4)
$$\begin{array}{r} 49 \\ -\ 37 \\ \hline \end{array}$$

(6)
$$\begin{array}{r} 56 \\ -\ 54 \\ \hline \end{array}$$

3 Calculate.

4 points per question

(1)
$$\begin{array}{r} 67 \\ -\ 14 \\ \hline \end{array}$$

(4)
$$\begin{array}{r} 75 \\ -\ 43 \\ \hline \end{array}$$

(7)
$$\begin{array}{r} 83 \\ -\ 50 \\ \hline \end{array}$$

(10)
$$\begin{array}{r} 65 \\ -\ 60 \\ \hline \end{array}$$

(2)
$$\begin{array}{r} 67 \\ -\ 25 \\ \hline \end{array}$$

(5)
$$\begin{array}{r} 75 \\ -\ 23 \\ \hline \end{array}$$

(8)
$$\begin{array}{r} 83 \\ -\ 62 \\ \hline \end{array}$$

(11)
$$\begin{array}{r} 95 \\ -\ \ 3 \\ \hline \end{array}$$

(3)
$$\begin{array}{r} 67 \\ -\ 36 \\ \hline \end{array}$$

(6)
$$\begin{array}{r} 75 \\ -\ 34 \\ \hline \end{array}$$

(9)
$$\begin{array}{r} 30 \\ -\ 10 \\ \hline \end{array}$$

(12)
$$\begin{array}{r} 92 \\ -\ \ 2 \\ \hline \end{array}$$

Subtraction in Vertical Form
2-Digits - 2-Digits 2

Date / /

Score /100

Review STEP 43

Calculate.

(1)
```
  1 3
- 1 1
```

(2)
```
  2 3
- 1 2
```

(3)
```
  3 4
- 2 4
```

(4)
```
  7 1
- 4 0
```

1 Calculate.

3 points per question

Example ● How to calculate 64 − 26

```
  5
  6 4
- 2 6
    8
```
The ones place

Carry over 1 from the tens place.
14 − 6 = 8

⇒

```
  5
  6 4
- 2 6
  3 8
```
The tens place

6 minus carried over 1 is 5.
5 − 2 = 3

(1)
```
  3 6
- 1 8
```

(3)
```
  4 4
- 2 7
```

(5)
```
  5 3
- 1 8
```

(7)
```
  6 2
- 2 5
```

(2)
```
  3 6
- 1 9
```

(4)
```
  4 4
- 2 8
```

(6)
```
  5 4
- 1 9
```

(8)
```
  6 3
- 2 4
```

2 Calculate.

4 points per question

(1)
$$\begin{array}{r} 42 \\ -\ 15 \\ \hline \end{array}$$

(3)
$$\begin{array}{r} 42 \\ -\ 29 \\ \hline \end{array}$$

(5)
$$\begin{array}{r} 50 \\ -\ 15 \\ \hline \end{array}$$

(7)
$$\begin{array}{r} 50 \\ -\ 27 \\ \hline \end{array}$$

(2)
$$\begin{array}{r} 42 \\ -\ 25 \\ \hline \end{array}$$

(4)
$$\begin{array}{r} 44 \\ -\ 15 \\ \hline \end{array}$$

(6)
$$\begin{array}{r} 50 \\ -\ 25 \\ \hline \end{array}$$

3 Calculate.

4 points per question

(1)
$$\begin{array}{r} 63 \\ -\ 17 \\ \hline \end{array}$$

(4)
$$\begin{array}{r} 64 \\ -\ 45 \\ \hline \end{array}$$

(7)
$$\begin{array}{r} 72 \\ -\ 45 \\ \hline \end{array}$$

(10)
$$\begin{array}{r} 52 \\ -\ 37 \\ \hline \end{array}$$

(2)
$$\begin{array}{r} 63 \\ -\ 28 \\ \hline \end{array}$$

(5)
$$\begin{array}{r} 72 \\ -\ 23 \\ \hline \end{array}$$

(8)
$$\begin{array}{r} 75 \\ -\ 57 \\ \hline \end{array}$$

(11)
$$\begin{array}{r} 80 \\ -\ 65 \\ \hline \end{array}$$

(3)
$$\begin{array}{r} 63 \\ -\ 36 \\ \hline \end{array}$$

(6)
$$\begin{array}{r} 72 \\ -\ 36 \\ \hline \end{array}$$

(9)
$$\begin{array}{r} 52 \\ -\ 24 \\ \hline \end{array}$$

(12)
$$\begin{array}{r} 80 \\ -\ 57 \\ \hline \end{array}$$

Subtraction in Vertical Form
2-Digits - 2-Digits 3

Date / /

Score /100

Review STEP 44

Calculate.

(1)
```
  66
- 48
```

(2)
```
  74
- 37
```

(3)
```
  31
- 12
```

(4)
```
  73
- 45
```

---

1 Calculate.

3 points per question

Example ● How to calculate 34 – 26

```
  ²34
- 26
   8
```

| The ones place | 14 − 6 = 8 |

When you subtract the numbers in the ones place it makes the numbers in the tens place the same.

The tens place is 0.

2 − 2 = 0

(1)
```
  24
-  7
```

(3)
```
  36
- 28
```

(5)
```
  36
- 27
```

(7)
```
  25
- 19
```

(2)
```
  24
- 17
```
□

(4)
```
  36
- 29
```

(6)
```
  24
- 17
```

(8)
```
  21
- 16
```

2 Calculate.

4 points per question

(1)
```
   4 3
 − 3 7
```

(2)
```
   4 3
 − 3 9
```

(3)
```
   4 3
 − 3 6
```

(4)
```
   4 3
 − 3 8
```

(5)
```
   5 3
 − 4 6
```

(6)
```
   5 2
 − 4 7
```

(7)
```
   5 1
 − 4 8
```

3 Calculate.

4 points per question

(1)
```
   6 5
 − 5 7
```

(2)
```
   6 5
 − 5 6
```

(3)
```
   6 5
 − 5 8
```

(4)
```
   6 4
 − 5 7
```

(5)
```
   7 5
 − 6 7
```

(6)
```
   5 5
 − 4 8
```

(7)
```
   7 5
 − 6 9
```

(8)
```
   7 3
 − 6 9
```

(9)
```
   8 4
 − 7 5
```

(10)
```
   8 3
 − 7 4
```

(11)
```
   8 4
 − 7 8
```

(12)
```
   9 2
 − 8 6
```

Subtraction in Vertical Form

Date / /

Score /100

Review STEP 43 **Calculate.**

2 points per question

(1)
```
   16
-  13
```

(3)
```
   19
-  13
```

(5)
```
   29
-  27
```

(7)
```
   39
-  18
```

(2)
```
   17
-  14
```

(4)
```
   18
-  16
```

(6)
```
   47
-  23
```

(8)
```
   26
-   6
```

Review STEP 44 **Calculate.**

2 points per question

(1)
```
   34
-  18
```

(4)
```
   42
-  29
```

(7)
```
   52
-  35
```

(10)
```
   62
-  28
```

(2)
```
   33
-  17
```

(5)
```
   51
-  25
```

(8)
```
   62
-  26
```

(11)
```
   51
-  18
```

(3)
```
   44
-  15
```

(6)
```
   45
-  16
```

(9)
```
   72
-  35
```

(12)
```
   31
-  14
```

Review STEP 44 **Calculate.** 3 points per question

(1)
$$\begin{array}{r} 60 \\ -\ 34 \\ \hline \end{array}$$

(3)
$$\begin{array}{r} 80 \\ -\ 33 \\ \hline \end{array}$$

(5)
$$\begin{array}{r} 60 \\ -\ 45 \\ \hline \end{array}$$

(7)
$$\begin{array}{r} 60 \\ -\ 34 \\ \hline \end{array}$$

(2)
$$\begin{array}{r} 70 \\ -\ 26 \\ \hline \end{array}$$

(4)
$$\begin{array}{r} 70 \\ -\ 38 \\ \hline \end{array}$$

(6)
$$\begin{array}{r} 80 \\ -\ 37 \\ \hline \end{array}$$

(8)
$$\begin{array}{r} 50 \\ -\ 23 \\ \hline \end{array}$$

Review STEP 45 **Calculate.** 3 points per question

(1)
$$\begin{array}{r} 27 \\ -\ 18 \\ \hline \end{array}$$

(4)
$$\begin{array}{r} 42 \\ -\ 37 \\ \hline \end{array}$$

(7)
$$\begin{array}{r} 25 \\ -\ 19 \\ \hline \end{array}$$

(10)
$$\begin{array}{r} 74 \\ -\ 69 \\ \hline \end{array}$$

(2)
$$\begin{array}{r} 27 \\ -\ 19 \\ \hline \end{array}$$

(5)
$$\begin{array}{r} 54 \\ -\ 49 \\ \hline \end{array}$$

(8)
$$\begin{array}{r} 35 \\ -\ 26 \\ \hline \end{array}$$

(11)
$$\begin{array}{r} 62 \\ -\ 58 \\ \hline \end{array}$$

(3)
$$\begin{array}{r} 42 \\ -\ 39 \\ \hline \end{array}$$

(6)
$$\begin{array}{r} 54 \\ -\ 47 \\ \hline \end{array}$$

(9)
$$\begin{array}{r} 74 \\ -\ 67 \\ \hline \end{array}$$

(12)
$$\begin{array}{r} 82 \\ -\ 79 \\ \hline \end{array}$$

STEP **46**

Subtraction in Vertical Form
....
3-Digit Subtraction I

Date / /

Score

/100

Review STEP 43 – STEP 45

Calculate.

(1)
```
   27
 - 18
```

(2)
```
   45
 - 16
```

(3)
```
   51
 - 25
```

(4)
```
   72
 - 38
```

1 Calculate.

3 points per question

Example
● How to calculate 143 − 28

```
   143
 -  28
     5
```

The ones place

Carry over 1 from the tens place.
13 − 8 = 5

⇨

```
   143
 -  28
   115
```

The tens place

4 minus carried over 1 is 3. 3 − 2 = 1

The hundreds place

Write 1.

(1)
```
   143
 -  26
```

(3)
```
   154
 -  49
```

(5)
```
   172
 -  26
```

(7)
```
   165
 -  47
```

(2)
```
   154
 -  27
```

(4)
```
   132
 -  14
```

(6)
```
   145
 -  18
```

(8)
```
   181
 -  34
```

2 Calculate.

4 points per question

(1)
$$\begin{array}{r} 257 \\ -\ 39 \\ \hline \end{array}$$

(3)
$$\begin{array}{r} 354 \\ -\ 27 \\ \hline \end{array}$$

(5)
$$\begin{array}{r} 442 \\ -\ 38 \\ \hline \end{array}$$

(7)
$$\begin{array}{r} 344 \\ -\ 36 \\ \hline \end{array}$$

(2)
$$\begin{array}{r} 257 \\ -\ 48 \\ \hline \end{array}$$

(4)
$$\begin{array}{r} 442 \\ -\ 16 \\ \hline \end{array}$$

(6)
$$\begin{array}{r} 244 \\ -\ 28 \\ \hline \end{array}$$

(8)
$$\begin{array}{r} 444 \\ -\ 29 \\ \hline \end{array}$$

3 Calculate.

4 points per question

(1)
$$\begin{array}{r} 153 \\ -\ 36 \\ \hline \end{array}$$

(4)
$$\begin{array}{r} 456 \\ -\ 49 \\ \hline \end{array}$$

(7)
$$\begin{array}{r} 110 \\ -\ 4 \\ \hline \end{array}$$

(10)
$$\begin{array}{r} 733 \\ -\ 28 \\ \hline \end{array}$$

(2)
$$\begin{array}{r} 253 \\ -\ 18 \\ \hline \end{array}$$

(5)
$$\begin{array}{r} 128 \\ -\ 19 \\ \hline \end{array}$$

(8)
$$\begin{array}{r} 540 \\ -\ 17 \\ \hline \end{array}$$

(11)
$$\begin{array}{r} 842 \\ -\ 23 \\ \hline \end{array}$$

(3)
$$\begin{array}{r} 356 \\ -\ 37 \\ \hline \end{array}$$

(6)
$$\begin{array}{r} 245 \\ -\ 26 \\ \hline \end{array}$$

(9)
$$\begin{array}{r} 663 \\ -\ 45 \\ \hline \end{array}$$

Subtraction in Vertical Form
3-Digit Subtraction 2

Review STEP 46

Calculate.

(1)
```
  240
-  36
```

(2)
```
  383
-  65
```

(3)
```
  764
-  46
```

(4)
```
  574
-  27
```

1 Calculate.

3 points per question

Example ● How to calculate 473 − 125

```
  473
- 125
    8
```
The ones place

Carry over 1 from the tens place.
13 − 5 = 8

⇨

```
  473
- 125
  348
```
The tens place

7 minus carried over 1 is 6. 6 − 2 = 4

The hundreds place

4 − 1 = 3

(1)
```
  357
- 147
```

(3)
```
  230
- 115
```

(5)
```
  473
- 127
```

(7)
```
  543
- 125
```

(2)
```
  357
- 148
```

(4)
```
  232
- 117
```

(6)
```
  473
- 148
```

(8)
```
  543
- 139
```

STEP 1-14
Mental Math
Addition

STEP 15-23
Mental Math
Subtraction

STEP 24-33
2-Digits Additon

STEP 34-42
3-Digits Additon

STEP 43-54
Subtraction in
Vertical Form

② Calculate.

4 points per question

(1)
$$\begin{array}{r} 534 \\ -\ 315 \\ \hline \end{array}$$

(3)
$$\begin{array}{r} 473 \\ -\ 235 \\ \hline \end{array}$$

(5)
$$\begin{array}{r} 645 \\ -\ 239 \\ \hline \end{array}$$

(7)
$$\begin{array}{r} 857 \\ -\ 248 \\ \hline \end{array}$$

(2)
$$\begin{array}{r} 534 \\ -\ 226 \\ \hline \end{array}$$

(4)
$$\begin{array}{r} 470 \\ -\ 168 \\ \hline \end{array}$$

(6)
$$\begin{array}{r} 645 \\ -\ 437 \\ \hline \end{array}$$

③ Calculate.

4 points per question

(1)
$$\begin{array}{r} 242 \\ -\ 125 \\ \hline \end{array}$$

(4)
$$\begin{array}{r} 535 \\ -\ 218 \\ \hline \end{array}$$

(7)
$$\begin{array}{r} 665 \\ -\ 136 \\ \hline \end{array}$$

(10)
$$\begin{array}{r} 535 \\ -\ 418 \\ \hline \end{array}$$

(2)
$$\begin{array}{r} 283 \\ -\ 126 \\ \hline \end{array}$$

(5)
$$\begin{array}{r} 465 \\ -\ 239 \\ \hline \end{array}$$

(8)
$$\begin{array}{r} 666 \\ -\ 237 \\ \hline \end{array}$$

(11)
$$\begin{array}{r} 250 \\ -\ 225 \\ \hline \end{array}$$

(3)
$$\begin{array}{r} 374 \\ -\ 159 \\ \hline \end{array}$$

(6)
$$\begin{array}{r} 465 \\ -\ 346 \\ \hline \end{array}$$

(9)
$$\begin{array}{r} 535 \\ -\ 317 \\ \hline \end{array}$$

(12)
$$\begin{array}{r} 377 \\ -\ 349 \\ \hline \end{array}$$

Review STEP 47

Calculate.

(1)
$$\begin{array}{r} 271 \\ -124 \\ \hline \end{array}$$

(2)
$$\begin{array}{r} 365 \\ -126 \\ \hline \end{array}$$

(3)
$$\begin{array}{r} 452 \\ -213 \\ \hline \end{array}$$

(4)
$$\begin{array}{r} 682 \\ -577 \\ \hline \end{array}$$

1 Calculate.

3 points per question

Example ● How to calculate $129 - 73$

$$\begin{array}{r} 129 \\ -73 \\ \hline 6 \end{array}$$
Calculate in the ones place.
$9 - 3 = 6$

⟹

$$\begin{array}{r} \overset{10}{1}29 \\ -73 \\ \hline 56 \end{array}$$

| The tens place |
Carry over 1 from the hundreds place.
$12 - 7 = 5$

(1)
$$\begin{array}{r} 150 \\ -10 \\ \hline \square\square\square \end{array}$$

(2)
$$\begin{array}{r} 120 \\ -50 \\ \hline \square\square \end{array}$$

(3)
$$\begin{array}{r} 120 \\ -70 \\ \hline \end{array}$$

(4)
$$\begin{array}{r} 140 \\ -50 \\ \hline \end{array}$$

(5)
$$\begin{array}{r} 128 \\ -40 \\ \hline \end{array}$$

(6)
$$\begin{array}{r} 136 \\ -40 \\ \hline \end{array}$$

(7)
$$\begin{array}{r} 127 \\ -43 \\ \hline \end{array}$$

(8)
$$\begin{array}{r} 135 \\ -43 \\ \hline \end{array}$$

2 Calculate.

4 points per question

(1)
```
  128
-  42
```

(3)
```
  128
-  65
```

(5)
```
  145
-  63
```

(7)
```
  135
-  83
```

(2)
```
  128
-  53
```

(4)
```
  128
-  81
```

(6)
```
  145
-  74
```

3 Calculate.

4 points per question

(1)
```
  136
-  53
```

(4)
```
  145
-  63
```

(7)
```
  167
-  95
```

(10)
```
  126
-  46
```

(2)
```
  136
-  74
```

(5)
```
  167
-  74
```

(8)
```
  152
-  81
```

(11)
```
  129
-  51
```

(3)
```
  145
-  52
```

(6)
```
  167
-  83
```

(9)
```
  142
-  62
```

(12)
```
  117
-  33
```

Subtraction in Vertical Form

3-Digit Subtraction 4

Date / /

Score /100

Review STEP 48

Calculate.

(1)
```
  1 1 0
-   8 0
```

(2)
```
  1 2 8
-   4 1
```

(3)
```
  1 4 5
-   5 3
```

(4)
```
  1 4 5
-   7 2
```

1 Calculate.

3 points per question

Example ● How to calculate 345 − 152

```
  3 4 5
- 1 5 2
  1 9 3
```

The ones place	$5 - 2 = 3$
The tens place	Carry over 1 from the hundreds place. $14 - 5 = 9$
The hundreds place	3 minus carried over 1 is 2. $2 - 1 = 1$

(1)
```
  3 2 4
- 1 6 2
```

(3)
```
  3 5 2
-   7 1
```

(5)
```
  4 2 5
- 1 8 2
```

(7)
```
  5 6 6
- 1 9 1
```

(2)
```
  3 2 4
- 1 5 2
```

(4)
```
  3 2 8
-   5 0
```

(6)
```
  4 4 7
- 1 5 3
```

(8)
```
  5 4 3
-   6 2
```

STEP 1-14
Mental Math Addition

STEP 15-23
Mental Math Subtraction

STEP 24-33
2-Digits Additon

STEP 34-42
3-Digits Additon

STEP 43-54
Subtraction in Vertical Form

2 Calculate.

4 points per question

(1)
$$\begin{array}{r} 634 \\ -\ 251 \\ \hline \end{array}$$

(3)
$$\begin{array}{r} 716 \\ -\ 251 \\ \hline \end{array}$$

(5)
$$\begin{array}{r} 827 \\ -\ 372 \\ \hline \end{array}$$

(7)
$$\begin{array}{r} 827 \\ -\ 663 \\ \hline \end{array}$$

(2)
$$\begin{array}{r} 634 \\ -\ 362 \\ \hline \end{array}$$

(4)
$$\begin{array}{r} 716 \\ -\ 320 \\ \hline \end{array}$$

(6)
$$\begin{array}{r} 827 \\ -\ 555 \\ \hline \end{array}$$

3 Calculate.

4 points per question

(1)
$$\begin{array}{r} 333 \\ -\ 151 \\ \hline \end{array}$$

(4)
$$\begin{array}{r} 628 \\ -\ 444 \\ \hline \end{array}$$

(7)
$$\begin{array}{r} 534 \\ -\ 372 \\ \hline \end{array}$$

(10)
$$\begin{array}{r} 432 \\ -\ 361 \\ \hline \end{array}$$

(2)
$$\begin{array}{r} 456 \\ -\ 62 \\ \hline \end{array}$$

(5)
$$\begin{array}{r} 357 \\ -\ 182 \\ \hline \end{array}$$

(8)
$$\begin{array}{r} 666 \\ -\ 86 \\ \hline \end{array}$$

(11)
$$\begin{array}{r} 319 \\ -\ 255 \\ \hline \end{array}$$

(3)
$$\begin{array}{r} 543 \\ -\ 372 \\ \hline \end{array}$$

(6)
$$\begin{array}{r} 465 \\ -\ 283 \\ \hline \end{array}$$

(9)
$$\begin{array}{r} 432 \\ -\ 260 \\ \hline \end{array}$$

(12)
$$\begin{array}{r} 247 \\ -\ 155 \\ \hline \end{array}$$

Review STEP 49

Calculate.

(1)
```
  4 1 2
- 1 8 0
```

(2)
```
  3 2 8
- 1 4 1
```

(3)
```
  5 4 1
- 2 7 1
```

(4)
```
  6 2 7
- 3 4 5
```

1 Calculate.

3 points per question

Example ● How to calculate 125 − 86

The ones place

Carry over 1 from the tens place.
15 − 6 = 9

⟹

The tens place

Carry over 1 from the hundreds place.
11 − 8 = 3

In this problem you have to carry over from both the tens place and the hundreds place.

(1)
```
  1 2 4
-   4 8
  □ □
```

(3)
```
  1 2 0
-   2 5
```

(5)
```
  1 2 6
-   2 9
```

(7)
```
  1 4 0
-   6 8
```

(2)
```
  1 2 4
-   5 8
```

(4)
```
  1 3 1
-   4 2
```

(6)
```
  1 2 3
-   9 6
```

(8)
```
  1 5 2
-   8 7
```

STEP 1-14
Mental Math
Addition

STEP 15-23
Mental Math
Subtraction

STEP 24-33
2-Digits Additon

STEP 34-42
3-Digits Additon

STEP 43-54
Subtraction in
Vertical Form

2 Calculate.

4 points per question

(1)
$$\begin{array}{r} 257 \\ -39 \\ \hline \end{array}$$

(3)
$$\begin{array}{r} 354 \\ -58 \\ \hline \end{array}$$

(5)
$$\begin{array}{r} 443 \\ -88 \\ \hline \end{array}$$

(7)
$$\begin{array}{r} 761 \\ -95 \\ \hline \end{array}$$

(2)
$$\begin{array}{r} 354 \\ -48 \\ \hline \end{array}$$

(4)
$$\begin{array}{r} 443 \\ -67 \\ \hline \end{array}$$

(6)
$$\begin{array}{r} 234 \\ -57 \\ \hline \end{array}$$

3 Calculate.

4 points per question

(1)
$$\begin{array}{r} 244 \\ -15 \\ \hline \end{array}$$

(4)
$$\begin{array}{r} 344 \\ -68 \\ \hline \end{array}$$

(7)
$$\begin{array}{r} 253 \\ -74 \\ \hline \end{array}$$

(10)
$$\begin{array}{r} 356 \\ -79 \\ \hline \end{array}$$

(2)
$$\begin{array}{r} 244 \\ -55 \\ \hline \end{array}$$

(5)
$$\begin{array}{r} 210 \\ -48 \\ \hline \end{array}$$

(8)
$$\begin{array}{r} 253 \\ -84 \\ \hline \end{array}$$

(11)
$$\begin{array}{r} 354 \\ -58 \\ \hline \end{array}$$

(3)
$$\begin{array}{r} 344 \\ -65 \\ \hline \end{array}$$

(6)
$$\begin{array}{r} 310 \\ -87 \\ \hline \end{array}$$

(9)
$$\begin{array}{r} 326 \\ -29 \\ \hline \end{array}$$

(12)
$$\begin{array}{r} 443 \\ -78 \\ \hline \end{array}$$

117

Subtraction in Vertical Form

3-Digit Subtraction 6

Date / /

Score /100

Review STEP 50

Calculate.

(1)
```
  220
-  36
```

(2)
```
  462
-  88
```

(3)
```
  535
-  57
```

(4)
```
  631
-  65
```

1 Calculate.

3 points per question

Example ● How to calculate 425 − 286

```
    1 10
  425        The ones place
- 286        Carry over 1 from
    9        the tens place.
             15 − 6 = 9
```
⇨
```
   3 1 10
  425        The tens place
- 286        Carry over 1 from the hundreds place. 11 − 8 = 3
  139
             The hundreds place
             4 minus carried over 1 is 3. 3 − 2 = 1
```

(1)
```
  342
- 164
```

(3)
```
  324
- 155
```

(5)
```
  617
- 159
```

(7)
```
  835
- 158
```

(2)
```
  356
- 168
```

(4)
```
  420
- 155
```

(6)
```
  617
- 269
```

(8)
```
  546
- 158
```

2 Calculate.

4 points per question

(1)
$$\begin{array}{r} 756 \\ -\ 287 \\ \hline \end{array}$$

(3)
$$\begin{array}{r} 674 \\ -\ 289 \\ \hline \end{array}$$

(5)
$$\begin{array}{r} 843 \\ -\ 367 \\ \hline \end{array}$$

(7)
$$\begin{array}{r} 953 \\ -\ 277 \\ \hline \end{array}$$

(2)
$$\begin{array}{r} 714 \\ -\ 256 \\ \hline \end{array}$$

(4)
$$\begin{array}{r} 453 \\ -\ 278 \\ \hline \end{array}$$

(6)
$$\begin{array}{r} 843 \\ -\ 687 \\ \hline \end{array}$$

3 Calculate.

4 points per question

(1)
$$\begin{array}{r} 865 \\ -\ 289 \\ \hline \end{array}$$

(4)
$$\begin{array}{r} 374 \\ -\ 197 \\ \hline \end{array}$$

(7)
$$\begin{array}{r} 843 \\ -\ 457 \\ \hline \end{array}$$

(10)
$$\begin{array}{r} 353 \\ -\ 278 \\ \hline \end{array}$$

(2)
$$\begin{array}{r} 873 \\ -\ 299 \\ \hline \end{array}$$

(5)
$$\begin{array}{r} 732 \\ -\ 256 \\ \hline \end{array}$$

(8)
$$\begin{array}{r} 843 \\ -\ 484 \\ \hline \end{array}$$

(11)
$$\begin{array}{r} 453 \\ -\ 397 \\ \hline \end{array}$$

(3)
$$\begin{array}{r} 378 \\ -\ 189 \\ \hline \end{array}$$

(6)
$$\begin{array}{r} 453 \\ -\ 268 \\ \hline \end{array}$$

(9)
$$\begin{array}{r} 714 \\ -\ 256 \\ \hline \end{array}$$

(12)
$$\begin{array}{r} 723 \\ -\ 329 \\ \hline \end{array}$$

3-Digit Subtraction 7

Review STEP 51

Calculate.

(1)
$$\begin{array}{r} 642 \\ -273 \\ \hline \end{array}$$

(2)
$$\begin{array}{r} 954 \\ -286 \\ \hline \end{array}$$

(3)
$$\begin{array}{r} 743 \\ -359 \\ \hline \end{array}$$

(4)
$$\begin{array}{r} 713 \\ -328 \\ \hline \end{array}$$

1 Calculate.

3 points per question

Example ● How to calculate 203 – 66

$$\begin{array}{r} 20\overset{9}{\cancel{\cancel{1}}}3 \\ -\ 66 \\ \hline 7 \end{array}$$

The ones place
It cannot be carried over from the hundreds place. So, calculate by carrying over 1 from the hundreds
13 – 6 = 7

⇨

$$\begin{array}{r} 2\overset{1}{\cancel{0}}\overset{9}{\cancel{\cancel{1}}}3 \\ -\ 66 \\ \hline 137 \end{array}$$

The tens place
10 minus carried over 1 is 9.
9 – 6 = 3
The hundreds place is 1.

(1)
$$\begin{array}{r} 203 \\ -\ 74 \\ \hline \end{array}$$

(3)
$$\begin{array}{r} 304 \\ -\ 86 \\ \hline \end{array}$$

(5)
$$\begin{array}{r} 403 \\ -\ 56 \\ \hline \end{array}$$

(7)
$$\begin{array}{r} 503 \\ -\ 68 \\ \hline \end{array}$$

(2)
$$\begin{array}{r} 303 \\ -\ 65 \\ \hline \end{array}$$

(4)
$$\begin{array}{r} 402 \\ -\ 95 \\ \hline \end{array}$$

(6)
$$\begin{array}{r} 403 \\ -\ 34 \\ \hline \end{array}$$

(8)
$$\begin{array}{r} 504 \\ -\ 79 \\ \hline \end{array}$$

2 Calculate.

4 points per question

(1)
$$107 - 9$$

(3)
$$103 - 18$$

(5)
$$200 - 8$$

(7)
$$405 - 9$$

(2)
$$101 - 6$$

(4)
$$104 - 25$$

(6)
$$300 - 57$$

3 Calculate.

4 points per question

(1)
$$403 - 127$$

(4)
$$704 - 546$$

(7)
$$803 - 245$$

(10)
$$306 - 107$$

(2)
$$403 - 147$$

(5)
$$720 - 134$$

(8)
$$803 - 346$$

(11)
$$604 - 328$$

(3)
$$704 - 324$$

(6)
$$720 - 502$$

(9)
$$300 - 168$$

(12)
$$800 - 264$$

Review STEP 51 STEP 52

Calculate.

(1)
```
  424
- 185
```

(2)
```
  821
- 373
```

(3)
```
  604
- 276
```

(4)
```
  503
- 209
```

1 Calculate.

5 points per question

Example ● How to calculate 1336 − 548

```
      2 10
  1 3 3 6
-   5 4 8
        8
```
The ones place
Carry over 1 from the tens place.

⇨

```
   10 10
   2  2
  1 3 3 6
-   5 4 8
  7 8 8
```
The tens place
Carry over 1 from the hundreds place.

The hundreds place
Carry over 1 from the thousands place.

(1)
```
  1234
-    6
```

(3)
```
  1234
-   46
```

(5)
```
  1546
-  273
```

(2)
```
  1234
-   16
```

(4)
```
  1546
-  219
```

(6)
```
  1546
-  288
```

2 Calculate.

5 points per question

(1)
```
  1358
-  548
```

(2)
```
  1336
-  342
```

(3)
```
  1242
-  361
```

(4)
```
  1248
-  576
```

(5)
```
  1534
-  577
```

(6)
```
  1732
-  855
```

3 Calculate.

5 points per question

(1)
```
  1000
-    6
```

(2)
```
  1401
-   16
```

(3)
```
  1502
-   54
```

(4)
```
  1865
-  597
```

(5)
```
  1731
-  693
```

(6)
```
  1256
-  398
```

(7)
```
  1073
-  368
```

(8)
```
  1051
-  592
```

STEP 54

Subtraction in Vertical Form
4-Digit Subtraction 2

Date   /   /

Score
        /100

Review STEP 53

Calculate.

(1)
```
  1358
-   79
```

(2)
```
  1623
-  751
```

(3)
```
  1548
-  399
```

1 Calculate.

5 points per question

Example ● How to calculate 3482 − 2541

```
  2 10
  3482
- 2541
   941
```

You can calculate the same way as 3-digit calculation even if the number of digits increases higher than 3-digit numbers.

(1)
```
  5000
- 3000
```

(3)
```
  3570
- 1230
```

(5)
```
  4685
- 3462
```

(2)
```
  6500
- 3200
```

(4)
```
  4780
- 2350
```

(6)
```
  5374
- 2168
```

2 Calculate.

5 points per question

(1)
$$\begin{array}{r} 4756 \\ -\ 2287 \\ \hline \end{array}$$

(3)
$$\begin{array}{r} 5673 \\ -\ 2384 \\ \hline \end{array}$$

(5)
$$\begin{array}{r} 4355 \\ -\ 1760 \\ \hline \end{array}$$

(2)
$$\begin{array}{r} 3734 \\ -\ 1256 \\ \hline \end{array}$$

(4)
$$\begin{array}{r} 7452 \\ -\ 1277 \\ \hline \end{array}$$

(6)
$$\begin{array}{r} 3734 \\ -\ 2489 \\ \hline \end{array}$$

3 Calculate.

5 points per question

(1)
$$\begin{array}{r} 4842 \\ -\ 1377 \\ \hline \end{array}$$

(4)
$$\begin{array}{r} 4704 \\ -\ 1267 \\ \hline \end{array}$$

(7)
$$\begin{array}{r} 5030 \\ -\ 2540 \\ \hline \end{array}$$

(2)
$$\begin{array}{r} 5284 \\ -\ 2655 \\ \hline \end{array}$$

(5)
$$\begin{array}{r} 6502 \\ -\ 2195 \\ \hline \end{array}$$

(8)
$$\begin{array}{r} 5030 \\ -\ 2170 \\ \hline \end{array}$$

(3)
$$\begin{array}{r} 3730 \\ -\ 1488 \\ \hline \end{array}$$

(6)
$$\begin{array}{r} 3000 \\ -\ 1680 \\ \hline \end{array}$$

TEST

Subtraction in Vertical-Form

Date / /

Score /100

Review STEP **46** STEP **47** **Calculate.**

2 points per question

(1)
```
  1 4 2
-   2 7
```

(3)
```
  2 5 2
- 1 3 8
```

(5)
```
  1 6 4
-   4 7
```

(7)
```
  3 4 1
- 1 3 5
```

(2)
```
  2 7 5
-   6 8
```

(4)
```
  4 7 3
- 2 3 9
```

(6)
```
  2 5 3
-   2 6
```

(8)
```
  7 6 8
- 5 1 9
```

Review STEP **48** STEP **49** **Calculate.**

3 points per question

(1)
```
  1 3 5
-   5 0
```

(4)
```
  1 4 6
-   9 3
```

(7)
```
  3 2 5
-   7 3
```

(10)
```
  6 2 9
- 4 6 8
```

(2)
```
  1 5 4
-   6 1
```

(5)
```
  1 6 2
-   8 2
```

(8)
```
  3 1 7
- 1 8 6
```

(3)
```
  1 1 7
-   4 3
```

(6)
```
  2 3 8
-   5 2
```

(9)
```
  5 4 6
- 1 9 3
```

STEP 1-14
Mental Math
Addition

STEP 15-23
Mental Math
Subtraction

STEP 24-33
2-Digits Additon

STEP 34-42
3-Digits Additon

STEP 43-54
Subtraction in
Vertical Form

Review STEP 50 – STEP 52 　Calculate.

3 points per question

(1)
$$\begin{array}{r} 145 \\ -\ 57 \\ \hline \end{array}$$

(4)
$$\begin{array}{r} 224 \\ -\ 159 \\ \hline \end{array}$$

(7)
$$\begin{array}{r} 652 \\ -\ 384 \\ \hline \end{array}$$

(10)
$$\begin{array}{r} 103 \\ -\ \ 8 \\ \hline \end{array}$$

(2)
$$\begin{array}{r} 134 \\ -\ 86 \\ \hline \end{array}$$

(5)
$$\begin{array}{r} 324 \\ -\ 155 \\ \hline \end{array}$$

(8)
$$\begin{array}{r} 821 \\ -\ 356 \\ \hline \end{array}$$

(11)
$$\begin{array}{r} 704 \\ -\ 259 \\ \hline \end{array}$$

(3)
$$\begin{array}{r} 223 \\ -\ 47 \\ \hline \end{array}$$

(6)
$$\begin{array}{r} 433 \\ -\ 185 \\ \hline \end{array}$$

(9)
$$\begin{array}{r} 202 \\ -\ 64 \\ \hline \end{array}$$

(12)
$$\begin{array}{r} 803 \\ -\ 207 \\ \hline \end{array}$$

Review STEP 53 STEP 54 　Calculate.

3 points per question

(1)
$$\begin{array}{r} 1443 \\ -\ \ 27 \\ \hline \end{array}$$

(3)
$$\begin{array}{r} 1325 \\ -\ 855 \\ \hline \end{array}$$

(5)
$$\begin{array}{r} 4734 \\ -\ 2378 \\ \hline \end{array}$$

(2)
$$\begin{array}{r} 1546 \\ -\ 627 \\ \hline \end{array}$$

(4)
$$\begin{array}{r} 6030 \\ -\ 4215 \\ \hline \end{array}$$

(6)
$$\begin{array}{r} 6175 \\ -\ 3846 \\ \hline \end{array}$$

Math Boosters

Grades 1-3 Addition & Subtraction

Answer Key

STEP 1 (P.4 · 5)

■ **Write the number that comes next.**

(1) 2   (3) 4   (5) 6
(2) 3   (4) 5   (6) 8

1
(1) 2   (4) 5   (7) 8
(2) 3   (5) 6   (8) 9
(3) 4   (6) 7   (9) 10

2
(1) 2   (4) 8   (7) 9
(2) 4   (5) 5   (8) 7
(3) 6   (6) 3   (9) 10

3
(1) 5   (7) 4   (13) 8
(2) 8   (8) 2   (14) 5
(3) 10   (9) 9   (15) 10
(4) 7   (10) 10 (16) 9
(5) 3   (11) 3
(6) 6   (12) 7

STEP 2 (P.6 · 7)

■ **Review of Step 1**

(1) 8   (3) 6   (5) 5
(2) 7   (4) 4   (6) 9

1
(1) 3   (4) 6   (7) 9
(2) 4   (5) 7   (8) 10
(3) 5   (6) 8   (9) 11

2
(1) 6   (4) 3   (7) 9
(2) 5   (5) 11 (8) 8
(3) 4   (6) 10 (9) 7

3
(1) 6   (7) 5   (13) 9
(2) 9   (8) 3   (14) 6
(3) 11   (9) 10   (15) 11
(4) 8   (10) 11 (16) 10
(5) 4   (11) 4
(6) 7   (12) 8

STEP 3 (P.8 · 9)

■ **Review of Step 2**

(1) 8   (3) 11 (5) 6
(2) 7   (4) 10 (6) 9

1
(1) 4   (4) 7   (7) 10
(2) 5   (5) 8   (8) 11
(3) 6   (6) 9   (9) 12

2
(1) 7   (4) 4   (7) 9
(2) 8   (5) 5   (8) 10
(3) 9   (6) 6   (9) 11

3
(1) 7   (7) 6   (13) 10
(2) 10   (8) 4   (14) 7
(3) 12   (9) 11   (15) 12
(4) 9   (10) 12 (16) 11
(5) 5   (11) 5
(6) 8   (12) 9

STEP 4 (P.10 · 11)

■ **Review of Step 2,3**

(1) 5   (3) 6   (5) 9
(2) 8   (4) 11 (6) 11

1
(1) 5   (4) 7   (7) 9
(2) 6   (5) 8   (8) 6
(3) 5   (6) 8   (9) 7

❷ (1) 9　　(4) 11　　(7) 11
　　(2) 9　　(5) 12　　(8) 12
　　(3) 10　 (6) 10　　(9) 13

❸ (1) 11　　(7) 9　　(13) 8
　　(2) 8　　 (8) 13　 (14) 6
　　(3) 12　　(9) 11　 (15) 13
　　(4) 10　　(10) 6　 (16) 7
　　(5) 7　　 (11) 9
　　(6) 5　　 (12) 12

STEP 5

■ Review of Step 4
　(1) 7　　(3) 8　　(5) 10
　(2) 6　　(4) 9　　(6) 12

❶ (1) 6　　(4) 8　　(7) 10
　　(2) 7　　(5) 9　　(8) 7
　　(3) 6　　(6) 9　　(9) 8

❷ (1) 10　　(4) 12　　(7) 12
　　(2) 10　　(5) 13　　(8) 13
　　(3) 11　　(6) 11　　(9) 14

❸ (1) 11　　(7) 14　　(13) 11
　　(2) 9　　 (8) 11　　(14) 14
　　(3) 7　　 (9) 13　　(15) 13
　　(4) 12　　(10) 8　　(16) 8
　　(5) 6　　 (11) 7
　　(6) 10　　(12) 12

TEST
(P.14 · 15)

■ Review of Step 1
　(1) 3　　(3) 6　　(5) 9
　(2) 4　　(4) 7　　(6) 10

■ Review of Step 2
　(1) 4　　(4) 8　　(7) 11
　(2) 5　　(5) 7　　(8) 10
　(3) 6　　(6) 9

■ Review of Step 3
　(1) 4　　(4) 8　　(7) 10
　(2) 6　　(5) 9　　(8) 12
　(3) 5　　(6) 7　　(9) 11

■ Review of Step 4
　(1) 6　　(4) 8　　(7) 10
　(2) 7　　(5) 9　　(8) 12
　(3) 5　　(6) 11　 (9) 13

■ Review of Step 5
　(1) 7　　(4) 8　　(7) 10
　(2) 6　　(5) 11　 (8) 13
　(3) 9　　(6) 12　 (9) 14

STEP 6
(P.16 · 17)

■ Review of Step 5
　(1) 8　　(3) 9　　(5) 12
　(2) 7　　(4) 10　 (6) 13

❶ (1) 7　　(4) 9　　(7) 11
　　(2) 8　　(5) 10　　(8) 8
　　(3) 7　　(6) 10　　(9) 9

❷ (1) 11　　(4) 13　　(7) 13
　　(2) 11　　(5) 14　　(8) 14
　　(3) 12　　(6) 12　　(9) 15

❸ (1) 10　　(7) 14　　(13) 7
　　(2) 13　　(8) 8　　 (14) 12
　　(3) 7　　 (9) 12　　(15) 14
　　(4) 11　　(10) 13　 (16) 9
　　(5) 15　　(11) 11
　　(6) 9　　 (12) 10

STEP 7 (P.18 · 19)

■ Review of Step 6

(1) 9	(3) 10	(5) 12
(2) 8	(4) 11	(6) 15

❶
(1) 8	(4) 10	(7) 12
(2) 9	(5) 11	(8) 9
(3) 8	(6) 11	(9) 10

❷
(1) 12	(4) 14	(7) 14
(2) 12	(5) 15	(8) 15
(3) 13	(6) 13	(9) 16

❸
(1) 11	(7) 15	(13) 15
(2) 14	(8) 9	(14) 16
(3) 8	(9) 11	(15) 14
(4) 12	(10) 12	(16) 9
(5) 16	(11) 10	
(6) 10	(12) 13	

STEP 8 (P.20 · 21)

■ Review of Step 7

(1) 10	(3) 11	(5) 14
(2) 9	(4) 12	(6) 16

❶
(1) 9	(4) 11	(7) 13
(2) 10	(5) 12	(8) 10
(3) 9	(6) 12	(9) 11

❷
(1) 13	(4) 14	(7) 16
(2) 13	(5) 15	(8) 16
(3) 14	(6) 15	(9) 17

❸
(1) 12	(7) 10	(13) 11
(2) 15	(8) 14	(14) 17
(3) 9	(9) 12	(15) 10
(4) 11	(10) 15	(16) 14
(5) 17	(11) 16	
(6) 16	(12) 13	

STEP 9 (P.22 · 23)

■ Review of Step 8

(1) 11	(3) 12	(5) 14
(2) 10	(4) 13	(6) 16

❶
(1) 10	(4) 12	(7) 14
(2) 11	(5) 13	(8) 11
(3) 10	(6) 13	(9) 12

❷
(1) 14	(4) 16	(7) 16
(2) 14	(5) 17	(8) 17
(3) 15	(6) 15	(9) 18

❸
(1) 13	(7) 15	(13) 17
(2) 16	(8) 14	(14) 18
(3) 14	(9) 18	(15) 13
(4) 12	(10) 12	(16) 15
(5) 11	(11) 10	
(6) 17	(12) 16	

TEST (P.24 · 25)

■ Review of Step 1,2

(1) 3	(4) 5	(7) 8
(2) 4	(5) 10	(8) 8
(3) 6	(6) 10	(9) 7

■ Review of Step 3,4

(1) 12	(4) 8	(7) 8
(2) 12	(5) 9	(8) 13
(3) 6	(6) 11	(9) 11

■ Review of Step 5

(1) 9	(4) 14	(7) 8
(2) 10	(5) 12	(8) 7
(3) 11	(6) 13	(9) 6

Review of Step 6,7

(1) 9   (5) 15   (9) 12
(2) 11   (6) 12   (10) 14
(3) 13   (7) 7   (11) 14
(4) 15   (8) 9

Review of Step 8,9

(1) 13   (5) 12   (9) 10
(2) 14   (6) 14   (10) 18
(3) 16   (7) 14   (11) 15
(4) 16   (8) 17   (12) 12

STEP 10 (P.26・27)

Review of Step 8,9

(1) 10   (3) 12   (5) 15
(2) 11   (4) 13   (6) 16

1 (1) 10   (4) 10   (7) 11
(2) 11   (5) 12   (8) 12
(3) 11   (6) 12   (9) 11

2 (1) 10   (5) 11   (9) 12
(2) 10   (6) 11   (10) 12
(3) 10   (7) 12   (11) 11
(4) 12   (8) 10

3 (1) 10   (5) 12   (9) 12
(2) 11   (6) 12   (10) 12
(3) 12   (7) 10
(4) 11   (8) 12

STEP 11 (P.28・29)

Review of Step 10

(1) 10   (3) 11   (5) 12
(2) 11   (4) 12   (6) 12

1 (1) 10   (4) 11   (7) 13
(2) 10   (5) 12   (8) 12
(3) 11   (6) 13   (9) 13

2 (1) 9   (5) 12   (9) 15
(2) 10   (6) 14   (10) 14
(3) 11   (7) 14   (11) 15
(4) 11   (8) 13

3 (1) 11   (5) 14   (9) 15
(2) 12   (6) 15   (10) 14
(3) 13   (7) 14
(4) 13   (8) 15

STEP 12 (P.30・31)

Review of Step 11

(1) 12   (3) 14   (5) 15
(2) 13   (4) 13   (6) 15

1 (1) 11   (4) 15   (7) 15
(2) 13   (5) 11   (8) 16
(3) 13   (6) 14   (9) 16

2 (1) 11   (5) 14   (9) 18
(2) 13   (6) 15   (10) 17
(3) 14   (7) 14   (11) 17
(4) 15   (8) 16

3 (1) 13   (5) 14   (9) 18
(2) 12   (6) 17   (10) 15
(3) 16   (7) 16
(4) 16   (8) 15

STEP 13 (P.32・33)

Review of Step 8,9

(1) 10   (3) 12   (5) 15
(2) 11   (4) 13   (6) 16

1 (1) 11   (4) 14   (7) 17
(2) 13   (5) 16   (8) 16
(3) 16   (6) 19   (9) 18

②
(1) 18   (4) 17   (7) 18
(2) 19   (5) 19   (8) 19
(3) 19   (6) 19   (9) 18

③
(1) 11   (7) 17   (13) 19
(2) 18   (8) 19   (14) 19
(3) 17   (9) 17   (15) 17
(4) 16   (10) 17   (16) 19
(5) 18   (11) 19
(6) 18   (12) 14

STEP 14

(P.34・35)

■ **Review of Step 13**
(1) 16   (3) 19   (5) 19
(2) 18   (4) 18   (6) 19

❶
(1) 20   (3) 60   (5) 40
(2) 30   (4) 70   (6) 50

❷
(1) 27   (4) 52   (7) 23
(2) 31   (5) 75   (8) 86
(3) 64   (6) 48

❸
(1) 80   (7) 20   (13) 40
(2) 90   (8) 30   (14) 38
(3) 70   (9) 40   (15) 80
(4) 80   (10) 26   (16) 40
(5) 57   (11) 80
(6) 90   (12) 70

TEST

(P.36・37)

■ **Review of Step 10**
(1) 11   (4) 10   (7) 12
(2) 10   (5) 10   (8) 11
(3) 12   (6) 11   (9) 12

■ **Review of Step 11**
(1) 10   (4) 13   (7) 15
(2) 11   (5) 14   (8) 15
(3) 12   (6) 14   (9) 15

■ **Review of Step 12**
(1) 12   (4) 14   (7) 18
(2) 13   (5) 15   (8) 16
(3) 13   (6) 16   (9) 17

■ **Review of Step 13**
(1) 14   (5) 19   (9) 17
(2) 16   (6) 15   (10) 16
(3) 15   (7) 19   (11) 18
(4) 16   (8) 19

■ **Review of Step 14**
(1) 70   (5) 40   (9) 38
(2) 80   (6) 70   (10) 90
(3) 90   (7) 90   (11) 70
(4) 56   (8) 80   (12) 83

STEP 15

(P.38・39)

■ **Write the number that comes before.**
(1) 1   (3) 3   (5) 5
(2) 2   (4) 4   (6) 7

❶
(1) 1   (4) 4   (7) 7
(2) 2   (5) 5   (8) 8
(3) 3   (6) 6   (9) 9

❷
(1) 3   (4) 0   (7) 7
(2) 2   (5) 9   (8) 6
(3) 1   (6) 8   (9) 4

3 (1) 1   (7) 4   (13) 9

(2) 3   (8) 6   (14) 4

(3) 5   (9) 8   (15) 6

(4) 7   (10) 0   (16) 8

(5) 9   (11) 5

(6) 2   (12) 7

STEP 16 (P.40・41)

■ Review of Step 15

(1) 3   (3) 4   (5) 8

(2) 2   (4) 6   (6) 9

1 (1) 1   (4) 4   (7) 7

(2) 2   (5) 5   (8) 8

(3) 3   (6) 6   (9) 9

2 (1) 1   (4) 6   (7) 3

(2) 2   (5) 5   (8) 8

(3) 7   (6) 0   (9) 9

3 (1) 1   (7) 6   (13) 1

(2) 5   (8) 4   (14) 0

(3) 2   (9) 2   (15) 7

(4) 8   (10) 3   (16) 8

(5) 7   (11) 6

(6) 9   (12) 4

STEP 17 (P.42・43)

■ Review of Step 16

(1) 2   (3) 5   (5) 8

(2) 3   (4) 6   (6) 9

1 (1) 1   (4) 4   (7) 7

(2) 2   (5) 5   (8) 8

(3) 3   (6) 6   (9) 9

2 (1) 1   (4) 6   (7) 9

(2) 0   (5) 5   (8) 4

(3) 7   (6) 8   (9) 3

3 (1) 2   (7) 5   (13) 4

(2) 1   (8) 8   (14) 7

(3) 3   (9) 6   (15) 0

(4) 6   (10) 9   (16) 1

(5) 4   (11) 3

(6) 7   (12) 2

STEP 18 (P.44・45)

■ Review of Step 17

(1) 2   (3) 3   (5) 7

(2) 4   (4) 6   (6) 9

1 (1) 3   (5) 3   (9) 3

(2) 2   (6) 1   (10) 2

(3) 1   (7) 0   (11) 1

(4) 4   (8) 4

2 (1) 4   (5) 5   (9) 5

(2) 5   (6) 3   (10) 3

(3) 3   (7) 4   (11) 6

(4) 1   (8) 2   (12) 4

3 (1) 5   (6) 7   (11) 7

(2) 2   (7) 5   (12) 6

(3) 7   (8) 6   (13) 5

(4) 3   (9) 4   (14) 8

(5) 5   (10) 9

STEP 19 (P.46・47)

■ Review of Step 18

(1) 3   (3) 3   (5) 5

(2) 1   (4) 4   (6) 3

1 (1) 4   (5) 4   (9) 5

(2) 2   (6) 6   (10) 8

(3) 3   (7) 4   (11) 6

(4) 5   (8) 6

②
- (1) 1
- (2) 2
- (3) 2
- (4) 1
- (5) 0
- (6) 2
- (7) 4
- (8) 1
- (9) 0
- (10) 4
- (11) 5
- (12) 2

③
- (1) 3
- (2) 4
- (3) 3
- (4) 2
- (5) 6
- (6) 1
- (7) 2
- (8) 4
- (9) 7
- (10) 1
- (11) 3
- (12) 2
- (13) 4
- (14) 4

STEP 20 (P.48・49)

■ Review of Step 19
- (1) 5
- (2) 7
- (3) 3
- (4) 4
- (5) 4
- (6) 2

❶
- (1) 8
- (2) 7
- (3) 5
- (4) 6
- (5) 6
- (6) 7
- (7) 9
- (8) 8
- (9) 7
- (10) 9
- (11) 8

❷
- (1) 4
- (2) 2
- (3) 3
- (4) 5
- (5) 3
- (6) 2
- (7) 5
- (8) 3
- (9) 4

❸
- (1) 5
- (2) 3
- (3) 2
- (4) 4
- (5) 4
- (6) 8
- (7) 3
- (8) 4
- (9) 6
- (10) 7
- (11) 5
- (12) 2
- (13) 3
- (14) 7
- (15) 9
- (16) 8
- (17) 6

STEP 21 (P.50・51)

■ Review of Step 20
- (1) 3
- (2) 5
- (3) 5
- (4) 7
- (5) 2
- (6) 2

❶
- (1) 7
- (2) 5
- (3) 6
- (4) 7
- (5) 7
- (6) 6
- (7) 8
- (8) 7
- (9) 9
- (10) 9
- (11) 8

❷
- (1) 3
- (2) 5
- (3) 3
- (4) 5
- (5) 6
- (6) 4
- (7) 5
- (8) 7
- (9) 6

❸
- (1) 9
- (2) 2
- (3) 6
- (4) 5
- (5) 8
- (6) 6
- (7) 8
- (8) 7
- (9) 8
- (10) 4
- (11) 7
- (12) 7
- (13) 5
- (14) 9
- (15) 6
- (16) 4
- (17) 9

STEP 22 (P.52・53)

■ Review of Step 21
- (1) 5
- (2) 9
- (3) 9
- (4) 4
- (5) 8
- (6) 8

❶
- (1) 7
- (2) 8
- (3) 9
- (4) 6
- (5) 7
- (6) 8
- (7) 6
- (8) 7
- (9) 8
- (10) 7
- (11) 8

❷
- (1) 7
- (2) 6
- (3) 7
- (4) 8
- (5) 9
- (6) 6
- (7) 9
- (8) 8
- (9) 7

❸
- (1) 5
- (2) 8
- (3) 5
- (4) 8
- (5) 8
- (6) 6
- (7) 6
- (8) 9
- (9) 6
- (10) 4
- (11) 7
- (12) 7
- (13) 9
- (14) 9
- (15) 7
- (16) 7
- (17) 4

STEP 23 (P.54・55)

■ Review of Step 22
(1) 6   (3) 6   (5) 8
(2) 8   (4) 7   (6) 4

❶
(1) 7   (5) 8   (9) 9
(2) 8   (6) 9   (10) 8
(3) 9   (7) 7   (11) 9
(4) 7   (8) 8

❷
(1) 8   (4) 8   (7) 9
(2) 7   (5) 7   (8) 8
(3) 6   (6) 9   (9) 9

❸
(1) 7   (7) 6   (13) 9
(2) 9   (8) 6   (14) 8
(3) 9   (9) 3   (15) 9
(4) 8   (10) 9   (16) 8
(5) 7   (11) 4   (17) 5
(6) 9   (12) 5

TEST (P.56・57)

■ Review of Step 15-17
(1) 2   (6) 9   (11) 9
(2) 6   (7) 2   (12) 5
(3) 9   (8) 7   (13) 8
(4) 2   (9) 6   (14) 0
(5) 6   (10) 0   (15) 4

■ Review of Step 18
(1) 3   (5) 2   (9) 5
(2) 4   (6) 5   (10) 3
(3) 6   (7) 2   (11) 1
(4) 4   (8) 0

■ Review of Step 19
(1) 5   (3) 1   (5) 4
(2) 3   (4) 2   (6) 1

■ Review of Step 20,21
(1) 2   (5) 9   (9) 7
(2) 3   (6) 4   (10) 9
(3) 8   (7) 5   (11) 8
(4) 5   (8) 6   (12) 6

■ Review of Step 22,23
(1) 6   (4) 8   (7) 8
(2) 7   (5) 9   (8) 9
(3) 8   (6) 9   (9) 7

STEP 24 (P.58・59)

■ Review of Step 13
(1) 11   (3) 14   (5) 16
(2) 12   (4) 14   (6) 18

❶
(1) 18   (3) 19   (5) 16   (7) 18
(2) 18   (4) 19   (6) 17   (8) 19

❷
(1) 16   (3) 19   (5) 18   (7) 18
(2) 17   (4) 17   (6) 18   (8) 19

❸
(1) 19   (4) 20   (7) 23   (10) 24
(2) 20   (5) 22   (8) 23   (11) 27
(3) 21   (6) 21   (9) 24

STEP 25 (P.60・61)

■ Review of Step 24
(1) 18   (2) 18   (3) 22   (4) 20

❶
(1) 24   (3) 46   (5) 38   (7) 49
(2) 35   (4) 57   (6) 45   (8) 53

❷
(1) 30   (3) 50   (5) 80   (7) 60
(2) 60   (4) 40   (6) 70

3 (1)
```
   63
+   9
   72
```
(5)
```
   42
+   8
   50
```

(2)
```
   77
+   8
   85
```
(6)
```
   84
+   6
   90
```

(3)
```
   38
+   9
   47
```
(7)
```
   75
+   7
   82
```

(4)
```
   44
+   8
   52
```
(8)
```
   56
+   8
   64
```

3 (1)
```
   56
+ 30
   86
```
(5)
```
   20
+   3
   23
```

(2)
```
   40
+ 32
   72
```
(6)
```
   60
+ 30
   90
```

(3)
```
   20
+ 70
   90
```
(7)
```
   60
+   8
   68
```

(4)
```
   40
+ 40
   80
```
(8)
```
   29
+ 30
   59
```

STEP 26

(P.62 · 63)

■ Review of Step 25
(1) 29　　(2) 36　　(3) 53　　(4) 73

1 (1) 26　　(3) 28　　(5) 37　　(7) 36
　　(2) 26　　(4) 29　　(6) 39　　(8) 39

2 (1) 43　　(3) 59　　(5) 55　　(7) 59
　　(2) 59　　(4) 66　　(6) 88

3 (1) 78　　(4) 76　　(7) 89　　(10) 96
　　(2) 99　　(5) 55　　(8) 84　　(11) 57
　　(3) 77　　(6) 39　　(9) 78　　(12) 97

STEP 27

(P.64 · 65)

■ Review of Step 26
(1) 38　　(2) 58　　(3) 79　　(4) 88

1 (1) 78　　(3) 74　　(5) 90　　(7) 80
　　(2) 56　　(4) 87　　(6) 80　　(8) 70

2 (1) 53　　(3) 73　　(5) 38　　(7) 57
　　(2) 55　　(4) 92　　(6) 64

STEP 28

(P.66 · 67)

■ Review of Step 27
(1) 77　　(2) 84　　(3) 90　　(4) 78

1 (1) 50　　(3) 70　　(5) 30　　(7) 50
　　(2) 60　　(4) 80　　(6) 40　　(8) 60

2 (1) 40　　(3) 60　　(5) 40　　(7) 60
　　(2) 50　　(4) 70　　(6) 51　　(8) 61

3 (1) 50　　(4) 90　　(7) 61　　(10) 60
　　(2) 70　　(5) 80　　(8) 72　　(11) 91
　　(3) 90　　(6) 71　　(9) 71

STEP 29 (P.68・69)

■ Review of Step 28
(1) 60　(2) 80　(3) 50　(4) 60

1 (1) 42　(3) 62　(5) 52　(7) 54
(2) 52　(4) 72　(6) 53　(8) 55

2 (1) 51　(3) 53　(5) 54　(7) 76
(2) 52　(4) 54　(6) 65　(8) 86

3 (1) 63　(4) 93　(7) 95　(10) 84
(2) 83　(5) 55　(8) 95　(11) 86
(3) 83　(6) 75　(9) 92

STEP 30 (P.70・71)

■ Review of Step 29
(1) 63　(2) 64　(3) 81　(4) 72

1 (1) 97　(3) 117　(5) 108　(7) 100
(2) 107　(4) 118　(6) 118　(8) 120

2 (1) 137　(4) 137　(7) 158　(10) 142
(2) 148　(5) 139　(8) 148　(11) 138
(3) 158　(6) 147　(9) 140

3 (1) 179　(3) 167　(5) 160　(7) 180
(2) 188　(4) 179　(6) 170　(8) 179

STEP 31 (P.72・73)

■ Review of Step 30
(1) 102　(2) 126　(3) 124　(4) 110

1 (1) 110　(3) 112　(5) 113　(7) 116
(2) 111　(4) 114　(6) 115　(8) 118

2 (1) 120　(3) 122　(5) 123　(7) 126
(2) 121　(4) 124　(6) 125　(8) 137

3 (1) 130　(4) 134　(7) 146　(10) 133
(2) 131　(5) 133　(8) 147　(11) 148
(3) 142　(6) 135　(9) 120

STEP 32 (P.74・75)

■ Review of Step 31
(1) 110　(2) 121　(3) 133　(4) 141

1 (1) 150　(3) 152　(5) 153　(7) 158
(2) 151　(4) 154　(6) 155　(8) 160

2 (1) 170　(4) 175　(7) 174　(10) 187
(2) 171　(5) 175　(8) 176　(11) 188
(3) 172　(6) 177　(9) 175

3 (1) 180　(3) 184　(5) 190　(7) 193
(2) 182　(4) 184　(6) 191　(8) 198

STEP 33 (P.76・77)

■ Review of Step 31,32
(1) 130　(2) 121　(3) 172　(4) 153

1 (1) 99　(3) 101　(5) 108　(7) 100
(2) 100　(4) 107　(6) 100　(8) 105

2 (1) 100　(3) 102　(5) 100　(7) 101
(2) 101　(4) 102　(6) 104　(8) 101

3 (1) 100　(4) 103　(7) 106　(10) 103
(2) 101　(5) 104　(8) 107　(11) 104
(3) 102　(6) 105　(9) 108

TEST

Review of Step 24,25

(1) 19 (4) 20 (7) 35 (10) 55
(2) 19 (5) 22 (8) 48
(3) 19 (6) 25 (9) 61

Review of Step 26,27

(1) 24 (3) 58 (5) 89 (7) 80
(2) 27 (4) 59 (6) 67 (8) 88

Review of Step 28,29

(1) 51 (3) 74 (5) 60 (7) 90
(2) 84 (4) 50 (6) 86 (8) 80

Review of Step 30-33

(1) 118 (4) 123 (7) 133 (10) 106
(2) 133 (5) 142 (8) 181
(3) 132 (6) 130 (9) 101

STEP 34

■ **Review of Step 30-33**

(1) 118 (2) 102 (3) 130 (4) 164

❶ (1) 122 (3) 143 (5) 140 (7) 170
(2) 131 (4) 153 (6) 160 (8) 180

❷ (1) 192 (4) 274 (7) 382 (10) 382
(2) 292 (5) 351 (8) 371 (11) 454
(3) 263 (6) 451 (9) 321

❸ (1) 173 (3) 292 (5) 892 (7) 791
(2) 491 (4) 384 (6) 986 (8) 992

STEP 35

■ **Review of Step 34**

(1) 152 (2) 171 (3) 361 (4) 362

❶ (1) 227 (3) 447 (5) 308 (7) 528
(2) 337 (4) 457 (6) 418 (8) 538

❷ (1) 209 (3) 429 (5) 217 (7) 537
(2) 319 (4) 539 (6) 327 (8) 637

❸ (1) 318 (4) 438 (7) 419 (10) 259
(2) 638 (5) 519 (8) 739 (11) 948
(3) 329 (6) 439 (9) 477

STEP 36

■ **Review of Step 35**

(1) 537 (2) 645 (3) 517 (4) 829

❶ (1) 222 (3) 443 (5) 331 (7) 421
(2) 424 (4) 477 (6) 311 (8) 431

❷ (1) 221 (3) 231 (5) 322 (7) 422
(2) 231 (4) 214 (6) 343 (8) 412

❸ (1) 445 (4) 510 (7) 860 (10) 625
(2) 431 (5) 663 (8) 876 (11) 924
(3) 533 (6) 744 (9) 924

STEP 37 (P.86・87)

■ Review of Step 36
- (1) 243
- (2) 351
- (3) 422
- (4) 723

❶
- (1) 300
- (3) 301
- (5) 401
- (7) 504
- (2) 300
- (4) 302
- (6) 402
- (8) 501

❷
- (1) 202
- (3) 300
- (5) 401
- (7) 501
- (2) 202
- (4) 300
- (6) 401
- (8) 600

❸
- (1) 501
- (4) 600
- (7) 403
- (10) 605
- (2) 803
- (5) 801
- (8) 906
- (11) 601
- (3) 500
- (6) 601
- (9) 904

STEP 38 (P.88・89)

■ Review of Step 37
- (1) 400
- (2) 501
- (3) 703
- (4) 601

❶
- (1) 1134
- (3) 1074
- (5) 1070
- (7) 1571
- (2) 1044
- (4) 1255
- (6) 1268
- (8) 1258

❷
- (1) 1031
- (3) 1133
- (5) 1032
- (7) 1566
- (2) 1043
- (4) 1142
- (6) 1151

❸
- (1) 1100
- (4) 1268
- (7) 1248
- (10) 1071
- (2) 1560
- (5) 1478
- (8) 1778
- (11) 1193
- (3) 1092
- (6) 1358
- (9) 1174
- (12) 1183

STEP 39 (P.90・91)

■ Review of Step 38
- (1) 1060
- (2) 1166
- (3) 1388
- (4) 1476

❶
- (1) 1215
- (3) 1218
- (5) 1215
- (7) 1213
- (2) 1254
- (4) 1308
- (6) 1315
- (8) 1317

❷
- (1) 1126
- (3) 1128
- (5) 1139
- (7) 1206
- (2) 1119
- (4) 1113
- (6) 1110
- (8) 1204

❸
- (1) 1117
- (4) 1546
- (7) 1334
- (10) 1307
- (2) 1218
- (5) 1628
- (8) 1535
- (11) 1105
- (3) 1436
- (6) 1109
- (9) 1538

STEP 40 (P.92・93)

■ Review of Step 39
- (1) 1125
- (2) 1424
- (3) 1254
- (4) 1535

❶
- (1) 1000
- (3) 1001
- (5) 1000
- (7) 1005
- (2) 1000
- (4) 1002
- (6) 1003
- (8) 1011

❷
- (1) 1100
- (4) 1224
- (7) 1134
- (10) 1133
- (2) 1223
- (5) 1002
- (8) 1323
- (11) 1131
- (3) 1130
- (6) 1122
- (9) 1151

❸
- (1) 1000
- (3) 1181
- (5) 1100
- (7) 1000
- (2) 1128
- (4) 1360
- (6) 1030
- (8) 1004

STEP 41 (P.94・95)

■ Review of Step 40
- (1) 1000
- (2) 1000
- (3) 1360
- (4) 1532

❶
- (1) 1100
- (3) 1630
- (5) 1579
- (2) 1870
- (4) 1888
- (6) 1589

❷
- (1) 2741
- (4) 2395
- (7) 1803
- (2) 2672
- (5) 2604
- (8) 4861
- (3) 3781
- (6) 1746

❸
- (1) 2200
- (3) 3214
- (5) 7227
- (2) 2692
- (4) 2264
- (6) 5029

STEP 42 (P.96・97)

■ Review of Step 41
- (1) 1845　(2) 3915　(3) 1821　(4) 3011

❶
- (1) 3000　(3) 4970　(5) 4659
- (2) 2400　(4) 3695　(6) 3599

❷
- (1) 3974　(4) 4882　(7) 4722
- (2) 5863　(5) 3678　(8) 3716
- (3) 5753　(6) 4663

❸
- (1) 3599　(3) 6523　(5) 4452
- (2) 4880　(4) 4655　(6) 4482

TEST (P.98・99)

Review of Step 34,35
- (1) 381　(3) 580　(5) 519　(7) 941
- (2) 376　(4) 892　(6) 618　(8) 607

Review of Step 36,37
- (1) 500　(3) 612　(5) 442　(7) 574
- (2) 412　(4) 623　(6) 833　(8) 910

Review of Step 38-40
- (1) 1270　(4) 1278　(7) 1223　(10) 1344
- (2) 1376　(5) 1016　(8) 1441　(11) 1044
- (3) 1164　(6) 1315　(9) 1121

Review of Step 41,42
- (1) 5155　(3) 5063　(5) 5441
- (2) 4211　(4) 3195　(6) 8790

STEP 43 (P.100・101)

■ Review of Step 23
- (1) 6　(3) 7　(5) 8
- (2) 9　(4) 9　(6) 8

❶
- (1) 3　(3) 13　(5) 11　(7) 11
- (2) 13　(4) 13　(6) 12　(8) 14

❷
- (1) 32　(3) 22　(5) 26　(7) 50
- (2) 32　(4) 12　(6) 2

❸
- (1) 53　(4) 32　(7) 33　(10) 5
- (2) 42　(5) 52　(8) 21　(11) 92
- (3) 31　(6) 41　(9) 20　(12) 90

STEP 44 (P.102・103)

■ Review of Step 43
- (1) 2　(2) 11　(3) 10　(4) 31

❶
- (1) 18　(3) 17　(5) 35　(7) 37
- (2) 17　(4) 16　(6) 35　(8) 39

❷
- (1) 27　(3) 13　(5) 35　(7) 23
- (2) 17　(4) 29　(6) 25

❸
- (1) 46　(4) 19　(7) 27　(10) 15
- (2) 35　(5) 49　(8) 18　(11) 15
- (3) 27　(6) 36　(9) 28　(12) 23

STEP 45 (P.104・105)

■ Review of Step 44
- (1) 18　(2) 37　(3) 19　(4) 28

❶
- (1) 17　(3) 8　(5) 9　(7) 6
- (2) 7　(4) 7　(6) 7　(8) 5

❷
- (1) 6　(3) 7　(5) 7　(7) 3
- (2) 4　(4) 5　(6) 5

3 (1) 8　　(4) 7　　(7) 6　　(10) 9
(2) 9　　(5) 8　　(8) 4　　(11) 6
(3) 7　　(6) 7　　(9) 9　　(12) 6

TEST

(P.106 · 107)

Review of Step 43
(1) 3　　(3) 6　　(5) 2　　(7) 21
(2) 3　　(4) 2　　(6) 24　　(8) 20

Review of Step 44
(1) 16　　(4) 13　　(7) 17　　(10) 34
(2) 16　　(5) 26　　(8) 36　　(11) 33
(3) 29　　(6) 29　　(9) 37　　(12) 17

Review of Step 44
(1) 26　　(3) 47　　(5) 15　　(7) 26
(2) 44　　(4) 32　　(6) 43　　(8) 27

Review of Step 45
(1) 9　　(4) 5　　(7) 6　　(10) 5
(2) 8　　(5) 5　　(8) 9　　(11) 4
(3) 3　　(6) 7　　(9) 7　　(12) 3

STEP 46

(P.108 · 109)

■ Review of Step 43-45
(1) 9　　(2) 29　　(3) 26　　(4) 34

1 (1) 117　　(3) 105　　(5) 146　　(7) 118
(2) 127　　(4) 118　　(6) 127　　(8) 147

2 (1) 218　　(3) 327　　(5) 404　　(7) 308
(2) 209　　(4) 426　　(6) 216　　(8) 415

3 (1) 117　　(4) 407　　(7) 106　　(10) 705
(2) 235　　(5) 109　　(8) 523　　(11) 819
(3) 319　　(6) 219　　(9) 618

STEP 47

(P.110 · 111)

■ Review of Step 46
(1) 204　　(2) 318　　(3) 718　　(4) 547

1 (1) 210　　(3) 115　　(5) 346　　(7) 418
(2) 209　　(4) 115　　(6) 325　　(8) 404

2 (1) 219　　(3) 238　　(5) 406　　(7) 609
(2) 308　　(4) 302　　(6) 208

3 (1) 117　　(4) 317　　(7) 529　　(10) 117
(2) 157　　(5) 226　　(8) 429　　(11) 25
(3) 215　　(6) 119　　(9) 218　　(12) 28

STEP 48

(P.112 · 113)

■ Review of Step 47
(1) 147　　(2) 239　　(3) 239　　(4) 105

1 (1) 140　　(3) 50　　(5) 88　　(7) 84
(2) 70　　(4) 90　　(6) 96　　(8) 92

2 (1) 86　　(3) 63　　(5) 82　　(7) 52
(2) 75　　(4) 47　　(6) 71

3 (1) 83　　(4) 82　　(7) 72　　(10) 80
(2) 62　　(5) 93　　(8) 71　　(11) 78
(3) 93　　(6) 84　　(9) 80　　(12) 84

STEP 49

(P.114 · 115)

■ Review of Step 48
(1) 30 　　(2) 87 　　(3) 92 　　(4) 73

❶
(1) 162 　　(3) 281 　　(5) 243 　　(7) 375
(2) 172 　　(4) 278 　　(6) 294 　　(8) 481

❷
(1) 383 　　(3) 465 　　(5) 455 　　(7) 164
(2) 272 　　(4) 396 　　(6) 272

❸
(1) 182 　　(4) 184 　　(7) 162 　　(10) 71
(2) 394 　　(5) 175 　　(8) 580 　　(11) 64
(3) 171 　　(6) 182 　　(9) 172 　　(12) 92

STEP 50

(P.116 · 117)

■ Review of Step 49
(1) 232 　　(2) 187 　　(3) 270 　　(4) 282

❶
(1) 76 　　(3) 95 　　(5) 97 　　(7) 72
(2) 66 　　(4) 89 　　(6) 27 　　(8) 65

❷
(1) 218 　　(3) 296 　　(5) 355 　　(7) 666
(2) 306 　　(4) 376 　　(6) 177

❸
(1) 229 　　(4) 276 　　(7) 179 　　(10) 277
(2) 189 　　(5) 162 　　(8) 169 　　(11) 296
(3) 279 　　(6) 223 　　(9) 297 　　(12) 365

STEP 51

(P.118 · 119)

■ Review of Step 50
(1) 184 　　(2) 374 　　(3) 478 　　(4) 566

❶
(1) 178 　　(3) 169 　　(5) 458 　　(7) 677
(2) 188 　　(4) 265 　　(6) 348 　　(8) 388

❷
(1) 469 　　(3) 385 　　(5) 476 　　(7) 676
(2) 458 　　(4) 175 　　(6) 156

❸
(1) 576 　　(4) 177 　　(7) 386 　　(10) 75
(2) 574 　　(5) 476 　　(8) 359 　　(11) 56
(3) 189 　　(6) 185 　　(9) 458 　　(12) 394

STEP 52

(P.120 · 121)

■ Review of Step 51
(1) 369 　　(2) 668 　　(3) 384 　　(4) 385

❶
(1) 129 　　(3) 218 　　(5) 347 　　(7) 435
(2) 238 　　(4) 307 　　(6) 369 　　(8) 425

❷
(1) 98 　　(3) 85 　　(5) 192 　　(7) 396
(2) 95 　　(4) 79 　　(6) 243

❸
(1) 276 　　(4) 158 　　(7) 558 　　(10) 199
(2) 256 　　(5) 586 　　(8) 457 　　(11) 276
(3) 380 　　(6) 218 　　(9) 132 　　(12) 536

STEP 53

(P.122 · 123)

■ Review of Step 51,52
(1) 239 　　(2) 448 　　(3) 328 　　(4) 294

❶
(1) 1228 　　(3) 1188 　　(5) 1273
(2) 1218 　　(4) 1327 　　(6) 1258

❷
(1) 810 　　(3) 881 　　(5) 957
(2) 994 　　(4) 672 　　(6) 877

❸
(1) 994 　　(4) 1268 　　(7) 705
(2) 1385 　　(5) 1038 　　(8) 459
(3) 1448 　　(6) 858

■ **Review of Step 53**

(1) 1279　　(2) 872　　(3) 1149

1 (1) 2000　　(3) 2340　　(5) 1223
　　(2) 3300　　(4) 2430　　(6) 3206

2 (1) 2469　　(3) 3289　　(5) 2595
　　(2) 2478　　(4) 6175　　(6) 1245

3 (1) 3465　　(4) 3437　　(7) 2490
　　(2) 2629　　(5) 4307　　(8) 2860
　　(3) 2242　　(6) 1320

TEST

Review of Step 46,47

(1) 115　(3) 114　(5) 117　(7) 206
(2) 207　(4) 234　(6) 227　(8) 249

Review of Step 48,49

(1) 85　(4) 53　(7) 252　(10) 161
(2) 93　(5) 80　(8) 131
(3) 74　(6) 186　(9) 353

Review of Step 50-52

(1) 88　(4) 65　(7) 268　(10) 95
(2) 48　(5) 169　(8) 465　(11) 445
(3) 176　(6) 248　(9) 138　(12) 596

Review of Step 53,54

(1) 1416　(3) 470　(5) 2356
(2) 919　(4) 1815　(6) 2329